Harnessing the Sun

A Comprehensive Guide to Solar Power for a
Sustainable Future

Maggie Soto

Harnessing the Sun

TABLE OF CONTENTS

Photovoltaic Cells: How They Work

Photovoltaic cells, often referred to as solar cells, are the fundamental building blocks of solar panels, converting sunlight into electricity through a fascinating process rooted in the principles of physics and chemistry. At the heart of this transformation lies the photovoltaic effect, a phenomenon discovered in the 19th century that has since revolutionized the way we harness energy from the sun.

The journey of a photon, a particle of light, begins as it travels from the sun to the Earth, carrying energy across the vast expanse of space. When these photons strike the surface of a photovoltaic cell, they encounter a carefully engineered structure designed to capture and convert their energy. The cell itself is typically composed of semiconductor materials, most commonly silicon, which have unique properties that make them ideal for this purpose.

Silicon, a widely abundant element, is used in its crystalline form to create the two layers of a photovoltaic cell: the n-type and p-type layers. These layers are created by doping silicon with other elements to alter its electrical properties. The n-type layer is infused with elements such as phosphorus, which adds extra electrons, while the p-type layer is doped with elements like boron, which creates spaces for electrons, known as "holes." When these two layers are placed together, they form a p-n junction, a critical component in the operation of the cell.

As photons penetrate the cell, their energy is absorbed by the silicon atoms, exciting electrons and freeing them from their atomic bonds. This process generates electron-hole pairs, which are essential for the creation of an electric current. The internal electric field at the p-n junction drives the electrons towards the n-type layer and the holes towards the p-type layer, creating a flow of electric charge. This movement of electrons constitutes an electric current, which can be harnessed and directed through an external circuit to power electrical devices.

The efficiency of a photovoltaic cell, or its ability to convert sunlight into usable electricity, is influenced by several factors. The quality of the semiconductor material, the design of the cell, and the presence of any impurities or defects all play a role in determining how effectively a cell can perform. Advances in technology have led to the development of various types of photovoltaic cells, each with its own unique characteristics and applications.

Monocrystalline silicon cells, known for their high efficiency and longevity, are made from a single crystal structure. They are easily recognizable by their uniform appearance and are often used in residential and commercial solar panels. Polycrystalline silicon cells, on the other hand, are composed of multiple crystal structures, giving them a distinctive, speckled look. While slightly less efficient than their monocrystalline counterparts, they are more cost-effective to produce and are widely used in large-scale solar installations.

Thin-film solar cells represent another category, characterized by their lightweight and flexible nature. These cells are made by depositing one or more thin layers of photovoltaic material onto a substrate, such as glass or metal. While they generally offer lower efficiency compared to crystalline silicon cells, their versatility and lower production costs make them suitable for a variety of applications, including building-integrated photovoltaics and portable solar devices.

Recent innovations in photovoltaic technology have introduced new materials and designs that promise to enhance the performance and affordability of solar cells. Perovskite solar cells, for example, have garnered significant attention due to their high efficiency and potential for low-cost production. These cells use a unique crystal structure that can be easily manufactured using solution-based processes, making them an attractive option for future solar technologies.

The integration of photovoltaic cells into everyday life extends beyond traditional solar panels. Building-integrated photovoltaics (BIPV) incorporate solar cells into the very fabric of buildings, such as windows, facades, and roofs, allowing for seamless energy generation without compromising aesthetics. This approach not only maximizes the use of available space but also contributes to the development of sustainable, energy-efficient buildings.

As the demand for renewable energy continues to grow, the future of photovoltaic technology looks promising. Researchers are exploring new materials, such as organic photovoltaics and quantum dots, which offer the potential for even greater efficiency and flexibility. Additionally, advancements in manufacturing techniques and economies of scale are driving down costs, making solar energy more accessible to individuals and businesses worldwide.

The environmental benefits of photovoltaic cells are undeniable. By harnessing the power of the sun, these cells provide a clean, renewable source of energy that reduces our reliance on fossil fuels and decreases greenhouse gas emissions. As technology continues to evolve, photovoltaic cells will play an increasingly vital role in the transition to a sustainable energy future, offering a glimpse of a world powered by the sun's abundant energy.

In the realm of energy generation, photovoltaic cells stand as a testament to human ingenuity and the relentless pursuit of innovation. Their ability to transform sunlight into electricity with remarkable efficiency and versatility underscores their significance in the global effort to combat climate change and secure a sustainable future for generations to come.

Types of Solar Panels and Their Applications

Solar panels have become a cornerstone of renewable energy solutions, offering a sustainable way to harness the sun's power. The diversity in solar panel types reflects the varied needs and applications across different sectors. Understanding these types and their specific applications can help individuals and businesses make informed decisions about integrating solar technology into their energy strategies.

Monocrystalline solar panels are among the most efficient and widely used types. These panels are made from a single, continuous crystal structure, which allows for a high degree of efficiency in converting sunlight into electricity. Their uniform appearance, often characterized by a deep black color, is a result of the pure silicon used in their construction. Monocrystalline panels are particularly well-suited for residential and commercial installations where space is at a premium, as their high efficiency means they can generate more power from a smaller footprint. This makes them an ideal choice for urban environments or areas with limited roof space.

Polycrystalline solar panels, on the other hand, are made from multiple silicon crystals melted together. This manufacturing process is less expensive than that of monocrystalline panels, making polycrystalline panels a more cost-effective option. While they are slightly less efficient, their blue hue and speckled appearance are distinctive. These panels are often used in large-scale solar farms where space is not a constraint, allowing for the installation of more panels to compensate for their lower efficiency. Their affordability makes them an attractive option for budget-conscious projects.

Thin-film solar panels represent a different approach to solar technology. These panels are created by depositing one or more thin layers of photovoltaic material onto a substrate, such as glass, plastic, or metal. The materials used can vary, including amorphous silicon, cadmium telluride, or copper indium gallium selenide. Thin-film panels are known for their flexibility and lightweight nature, which opens up a range of applications not possible with traditional rigid panels. They can be integrated into building materials, such as solar shingles or facades, and are ideal for portable solar solutions. Their lower efficiency is offset by their versatility and ease of installation, making them suitable for innovative architectural designs and off-grid applications.

Bifacial solar panels are an emerging technology that captures sunlight from both sides of the panel. This design allows them to harness reflected sunlight from surfaces such as rooftops or the ground, potentially increasing their energy output. Bifacial panels are particularly effective in environments with high albedo, where the ground reflects a significant amount of sunlight. They are often used in solar farms with reflective surfaces or in snowy regions where the ground naturally reflects sunlight. The dual-sided nature of these panels can lead to higher energy yields, making them an attractive option for maximizing efficiency.

Concentrated photovoltaic (CPV) panels take a different approach by using lenses or mirrors to focus sunlight onto high-efficiency solar cells. This concentration of sunlight allows CPV panels to achieve efficiencies far beyond those of traditional panels. However, they require direct sunlight and tracking systems to maintain optimal alignment with the sun, which can increase installation complexity and cost. CPV panels are best suited for areas with high direct sunlight and are often used in utility-scale solar power plants where land is abundant and solar tracking systems can be effectively implemented.

The choice of solar panel type is influenced by several factors, including geographic location, available space, budget, and specific energy needs. For residential applications, monocrystalline panels are often favored for their efficiency and aesthetic appeal, while polycrystalline panels offer a more budget-friendly alternative. In commercial settings, the decision may hinge on the balance between cost and efficiency, with thin-film panels providing flexibility for unique architectural designs. For large-scale solar farms, the choice between polycrystalline and bifacial panels may depend on land availability and environmental conditions.

Innovations in solar panel technology continue to expand the possibilities for solar energy applications. Building-integrated photovoltaics (BIPV) are gaining traction as a way to seamlessly incorporate solar panels into the design of buildings, turning roofs, windows, and facades into energy-generating surfaces. This approach not only enhances the aesthetic appeal of solar installations but also maximizes the use of available space, contributing to the development of energy-efficient buildings.

The environmental impact of solar panels is another important consideration. While the production of solar panels involves energy and resources, the long-term benefits of reduced greenhouse gas emissions and reliance on fossil fuels are significant. Advances in recycling technologies are also addressing concerns about the disposal of solar panels at the end of their lifecycle, further enhancing their sustainability credentials.

As solar technology continues to evolve, the range of applications for solar panels is expanding. From powering remote villages to integrating into urban infrastructure, solar panels are playing a crucial role in the transition to a sustainable energy future. By understanding the different types of solar panels and their applications, individuals and businesses can make informed decisions that align with their energy goals and contribute to a cleaner, more sustainable world.

Innovations in Solar Technology

Solar technology has undergone remarkable transformations over the past few decades, driven by the relentless pursuit of efficiency, cost-effectiveness, and sustainability. These innovations have not only expanded the potential applications of solar energy but have also made it more accessible to a broader audience. As the world continues to seek alternatives to fossil fuels, the advancements in solar technology stand as a beacon of hope for a cleaner, more sustainable future.

One of the most significant innovations in solar technology is the development of perovskite solar cells. Named after the mineral with a similar crystal structure, perovskite materials have shown tremendous promise due to their high efficiency and low production costs. Unlike traditional silicon-based solar cells, perovskite cells can be manufactured using solution-based processes, which are simpler and less energy-intensive. This has the potential to significantly reduce the cost of solar energy, making it more competitive with conventional energy sources. Moreover, perovskite cells can be made semi-transparent, opening up new possibilities for integration into windows and other building materials, thereby transforming urban landscapes into energy-generating environments.

Another groundbreaking innovation is the advent of bifacial solar panels. Unlike traditional panels that capture sunlight on one side, bifacial panels are designed to absorb light from both sides. This allows them to harness reflected sunlight from surfaces such as rooftops or the ground, potentially increasing their energy output by up to 30%. Bifacial panels are particularly effective in environments with high albedo, where the ground reflects a significant amount of sunlight. Their ability to generate more electricity from the same surface area makes them an attractive option for maximizing efficiency in solar farms and other large-scale installations.

Floating solar farms, or floatovoltaics, represent a novel approach to solar energy generation. By installing solar panels on bodies of water, such as reservoirs or lakes, these systems take advantage of unused space while also reducing water evaporation and algae growth. The cooling effect of the water can also enhance the efficiency of the panels, leading to higher energy yields. Floating solar farms are particularly beneficial in regions with limited land availability, providing a sustainable solution that does not compete with agricultural or urban land use.

The integration of solar technology into everyday objects is another area of innovation that is gaining momentum. Solar-powered wearables, such as watches and fitness trackers, are becoming increasingly popular, offering a convenient way to keep devices charged without the need for traditional power sources. Similarly, solar backpacks equipped with photovoltaic panels allow users to charge their electronic devices on the go, making them ideal for outdoor enthusiasts and travelers. These innovations highlight the versatility of solar technology and its potential to seamlessly integrate into our daily lives.

Energy storage solutions have also seen significant advancements, addressing one of the key challenges of solar energy: its intermittent nature. The development of more efficient and affordable battery technologies, such as lithium-ion and solid-state batteries, has made it possible to store solar energy for use during periods of low sunlight. This has paved the way for the creation of solar-plus-storage systems, which combine solar panels with battery storage to provide a reliable and continuous power supply. These systems are particularly valuable in off-grid and remote areas, where access to traditional power sources is limited.

The concept of solar roads is another innovative idea that is being explored. By embedding solar panels into road surfaces, it is possible to generate electricity while also providing a durable and functional roadway. Solar roads have the potential to power streetlights, traffic signals, and even electric vehicles, creating a self-sustaining infrastructure that reduces reliance on external power sources. While still in the experimental stage, solar roads represent a bold vision for the future of transportation and energy generation.

In the realm of agriculture, solar technology is being harnessed to improve efficiency and sustainability. Agrivoltaics, the practice of combining agriculture with solar energy production, allows for the dual use of land, maximizing its productivity. By installing solar panels above crops, farmers can generate electricity while also providing shade and reducing water evaporation. This approach not only increases the overall efficiency of land use but also offers economic benefits to farmers by diversifying their income streams.

The rise of smart solar technology is another exciting development. By integrating solar panels with smart grid systems and Internet of Things (IoT) devices, it is possible to optimize energy production and consumption in real-time. Smart solar systems can monitor weather conditions, track energy usage, and adjust panel orientation to maximize efficiency. This level of automation and control enhances the reliability and performance of solar installations, making them more attractive to consumers and businesses alike.

As solar technology continues to evolve, the potential for innovation remains vast. Researchers are exploring new materials, such as organic photovoltaics and quantum dots, which offer the promise of even greater efficiency and flexibility. Additionally, advancements in manufacturing techniques and economies of scale are driving down costs, making solar energy more accessible to individuals and businesses worldwide.

The environmental benefits of these innovations are profound. By reducing our reliance on fossil fuels and decreasing greenhouse gas emissions, solar technology is playing a crucial role in the global effort to combat climate change. As we continue to push the boundaries of what is possible, the future of solar technology looks brighter than ever, offering a glimpse of a world powered by the sun's abundant energy.

Efficiency and Performance Factors

Efficiency and performance are critical considerations when evaluating solar technology, as they directly impact the effectiveness and economic viability of solar energy systems. Understanding the factors that influence these aspects can help individuals and businesses optimize their solar installations and maximize their return on investment.

The efficiency of a solar panel refers to its ability to convert sunlight into usable electricity. This is typically expressed as a percentage, indicating the proportion of sunlight that is transformed into electrical energy. Several factors contribute to the efficiency of a solar panel, starting with the quality of the photovoltaic cells themselves. High-quality materials and precise manufacturing processes are essential for producing cells that can effectively capture and convert solar energy.

Temperature plays a significant role in the performance of solar panels. While it might seem counterintuitive, solar panels are less efficient at higher temperatures. As the temperature rises, the electrical resistance within the cells increases, leading to a decrease in their ability to generate electricity. This phenomenon is known as the temperature coefficient, and it varies among different types of panels. For instance, monocrystalline panels typically have a lower temperature coefficient compared to polycrystalline panels, making them more suitable for hot climates.

The angle and orientation of solar panels also significantly affect their performance. To maximize energy production, panels should be positioned to capture the maximum amount of sunlight throughout the day. This often involves angling the panels to match the latitude of the installation site and orienting them towards the equator. In the northern hemisphere, this means facing panels south, while in the southern hemisphere, they should face north. Tracking systems can further enhance performance by adjusting the panel's position to follow the sun's path across the sky, although these systems can add complexity and cost to the installation.

Shading is another critical factor that can dramatically reduce the efficiency of solar panels. Even partial shading from trees, buildings, or other obstructions can lead to significant drops in energy production. This is because most solar panels are composed of multiple cells connected in series, and shading a single cell can affect the performance of the entire panel. To mitigate this issue, microinverters or power optimizers can be used to allow each panel or cell to operate independently, minimizing the impact of shading on the overall system.

The cleanliness of solar panels is often overlooked but plays a vital role in maintaining their efficiency. Dust, dirt, and debris can accumulate on the surface of panels, blocking sunlight and reducing their ability to generate electricity. Regular cleaning and maintenance are essential to ensure that panels operate at peak performance. In areas with frequent rainfall, natural cleaning may suffice, but in arid regions, manual cleaning may be necessary to prevent significant efficiency losses.

The quality of the inverter, which converts the direct current (DC) produced by solar panels into alternating current (AC) for use in homes and businesses, is another crucial factor in overall system performance. High-quality inverters are more efficient and reliable, ensuring that the maximum amount of electricity generated by the panels is available for use. Inverter efficiency is typically around 95-98%, but choosing a reputable brand and model can make a difference in long-term performance and reliability.

Advancements in solar technology continue to push the boundaries of efficiency. Researchers are exploring new materials and cell designs that promise higher efficiencies and better performance under various conditions. For example, tandem solar cells, which combine multiple layers of photovoltaic materials, are being developed to capture a broader spectrum of sunlight, potentially increasing efficiency beyond the limits of traditional silicon-based cells.

The performance of a solar energy system is also influenced by external factors such as weather conditions and seasonal variations. Cloud cover, for instance, can significantly reduce the amount of sunlight reaching the panels, leading to lower energy production. However, modern solar systems are designed to handle these fluctuations, and energy storage solutions, such as batteries, can help provide a consistent power supply even during periods of low sunlight.

Monitoring and data analysis are becoming increasingly important in optimizing the performance of solar installations. By tracking energy production and identifying patterns or anomalies, system owners can make informed decisions about maintenance, upgrades, or adjustments to improve efficiency. Many modern solar systems come equipped with monitoring software that provides real-time data and insights, allowing users to maximize their energy output and savings.

The economic implications of efficiency and performance are significant. Higher efficiency panels may come with a higher upfront cost, but their ability to generate more electricity over their lifespan can lead to greater savings and a faster return on investment. Additionally, incentives and rebates offered by governments and utilities can further enhance the financial benefits of investing in high-performance solar technology.

As the demand for renewable energy continues to grow, the focus on efficiency and performance will remain a driving force in the development of solar technology. By understanding the factors that influence these aspects, individuals and businesses can make informed decisions that align with their energy goals and contribute to a more sustainable future. Solar technology offers a powerful solution to the challenges of energy production and environmental sustainability, and optimizing its efficiency and performance is key to unlocking its full potential.

Future Trends in Solar Technology

The landscape of solar technology is rapidly evolving, driven by the urgent need for sustainable energy solutions and the relentless pursuit of innovation. As we look to the future, several trends are poised to reshape the solar industry, offering new opportunities and challenges for individuals, businesses, and governments alike.

One of the most promising trends is the continued development and commercialization of perovskite solar cells. These cells have garnered significant attention due to their high efficiency and potential for low-cost production. Unlike traditional silicon-based cells, perovskite cells can be manufactured using simpler processes, which could dramatically reduce the cost of solar energy. Researchers are making strides in improving the stability and durability of perovskite cells, addressing concerns about their long-term performance. As these challenges are overcome, perovskite solar cells are expected to play a crucial role in expanding the accessibility and affordability of solar energy.

The integration of solar technology into everyday life is another trend gaining momentum. Building-integrated photovoltaics (BIPV) are becoming increasingly popular, allowing solar panels to be seamlessly incorporated into the design of buildings. This includes solar shingles, facades, and windows that generate electricity while maintaining aesthetic appeal. As urban areas continue to grow, BIPV offers a sustainable solution for maximizing energy generation without compromising architectural design. This trend is likely to accelerate as advancements in materials and manufacturing techniques make BIPV more cost-effective and efficient.

Energy storage solutions are also evolving, addressing one of the key limitations of solar energy: its intermittent nature. The development of more efficient and affordable battery technologies, such as lithium-ion and solid-state batteries, is enabling the widespread adoption of solar-plus-storage systems. These systems allow for the storage of excess solar energy generated during the day for use during periods of low sunlight, providing a reliable and continuous power supply. As battery technology continues to improve, we can expect to see greater integration of solar energy into the grid, enhancing its stability and resilience.

The rise of smart solar technology is transforming the way we manage and optimize solar energy systems. By integrating solar panels with smart grid systems and Internet of Things (IoT) devices, it is possible to monitor and control energy production and consumption in real-time. Smart solar systems can adjust panel orientation, track weather conditions, and optimize energy usage, maximizing efficiency and reducing costs. This level of automation and control is becoming increasingly important as the complexity of solar installations grows, offering new opportunities for innovation and efficiency.

Floating solar farms, or floatovoltaics, represent a novel approach to solar energy generation. By installing solar panels on bodies of water, such as reservoirs or lakes, these systems take advantage of unused space while also reducing water evaporation and algae growth. The cooling effect of the water can enhance the efficiency of the panels, leading to higher energy yields. Floating solar farms are particularly beneficial in regions with limited land availability, providing a sustainable solution that does not compete with agricultural or urban land use. As the technology matures, we can expect to see more widespread adoption of floatovoltaics, particularly in densely populated areas.

The concept of solar roads is another innovative idea that is being explored. By embedding solar panels into road surfaces, it is possible to generate electricity while also providing a durable and functional roadway. Solar roads have the potential to power streetlights, traffic signals, and even electric vehicles, creating a self-sustaining infrastructure that reduces reliance on external power sources. While still in the experimental stage, solar roads represent a bold vision for the future of transportation and energy generation, offering a glimpse of a world where every surface can contribute to energy production.

The environmental benefits of these future trends are profound. By reducing our reliance on fossil fuels and decreasing greenhouse gas emissions, solar technology is playing a crucial role in the global effort to combat climate change. As we continue to push the boundaries of what is possible, the future of solar technology looks brighter than ever, offering a glimpse of a world powered by the sun's abundant energy.

The economic implications of these trends are equally significant. As solar technology becomes more efficient and affordable, it is expected to drive down the cost of electricity, making it more competitive with traditional energy sources. This has the potential to spur economic growth and job creation, particularly in the renewable energy sector. Governments and businesses are increasingly recognizing the value of investing in solar technology, leading to greater support for research and development, as well as incentives for adoption.

The future of solar technology is not without its challenges. Issues such as the environmental impact of solar panel production, the need for improved recycling processes, and the integration of solar energy into existing infrastructure must be addressed. However, the potential benefits far outweigh these challenges, and continued innovation and collaboration will be key to overcoming them.

As we look to the future, it is clear that solar technology will play a central role in the transition to a sustainable energy future. By embracing these trends and investing in research and development, we can unlock the full potential of solar energy and create a cleaner, more sustainable world for generations to come. The journey towards a solar-powered future is just beginning, and the possibilities are as limitless as the sun itself.

Assessing Solar Potential for Homes

Harnessing solar energy for residential use is an increasingly attractive option for homeowners seeking to reduce their carbon footprint and energy bills. Assessing the solar potential of a home involves evaluating several key factors that determine the feasibility and efficiency of a solar installation. By understanding these elements, homeowners can make informed decisions about whether solar energy is a viable option for their property.

The first step in assessing solar potential is to evaluate the geographic location of the home. Solar energy production is highly dependent on the amount of sunlight a location receives, which varies based on latitude, climate, and weather patterns. Homes situated in regions with high solar insolation, such as the southwestern United States or southern Europe, are particularly well-suited for solar installations. However, even in less sunny areas, solar energy can still be a viable option, especially with advancements in solar technology that improve efficiency under diffuse light conditions.

The orientation and tilt of the roof are critical factors in determining solar potential. Ideally, solar panels should be installed on a south-facing roof in the northern hemisphere or a north-facing roof in the southern hemisphere to capture the maximum amount of sunlight throughout the day. The angle of the roof should also be considered, as it affects the panels' exposure to the sun. A tilt angle that matches the latitude of the location is generally optimal, but adjustable mounting systems can be used to fine-tune the angle for maximum efficiency.

Shading is another important consideration when assessing solar potential. Trees, buildings, and other obstructions can cast shadows on solar panels, significantly reducing their energy output. Conducting a shading analysis can help identify potential sources of shade and determine the best placement for solar panels. In some cases, trimming trees or choosing a different roof section for installation may be necessary to minimize shading and maximize energy production.

The size and condition of the roof also play a role in determining solar potential. A roof with ample space and a structurally sound foundation is ideal for supporting the weight of solar panels and associated equipment. It's important to assess the age and condition of the roof before installation, as older roofs may require repairs or replacement to ensure they can support the solar system for its expected lifespan. Additionally, the roof's material can impact the ease of installation and the type of mounting system used.

Financial considerations are a crucial aspect of assessing solar potential. The cost of solar panels, inverters, and installation can vary widely, and homeowners should evaluate their budget and financing options before proceeding. Many governments and utilities offer incentives, rebates, or tax credits to offset the initial cost of solar installations, making them more affordable for homeowners. It's essential to research and understand the available incentives in your area to take full advantage of these opportunities.

Energy consumption patterns should also be analyzed to determine the appropriate size of the solar system. By reviewing past utility bills and understanding daily and seasonal energy usage, homeowners can estimate the amount of electricity their solar system needs to generate to meet their needs. This analysis can help determine the number of panels required and whether additional energy storage solutions, such as batteries, are necessary to provide a consistent power supply.

The potential for net metering is another factor to consider when assessing solar potential. Net metering allows homeowners to sell excess electricity generated by their solar system back to the grid, providing a credit on their utility bill. This can significantly enhance the financial benefits of a solar installation, particularly in regions where net metering policies are favorable. It's important to research local net metering regulations and understand how they will impact the overall economics of the solar system.

Advancements in solar technology have expanded the range of options available to homeowners, making it easier to find a solution that fits their specific needs. For example, building-integrated photovoltaics (BIPV) offer an alternative to traditional solar panels by incorporating solar cells directly into roofing materials, windows, or facades. This approach can be particularly appealing for homeowners who prioritize aesthetics or have limited roof space.

The environmental impact of a solar installation is another consideration for homeowners assessing solar potential. By reducing reliance on fossil fuels and decreasing greenhouse gas emissions, solar energy contributes to a cleaner and more sustainable environment. Understanding the environmental benefits of solar energy can provide additional motivation for homeowners to invest in solar technology and contribute to the global effort to combat climate change.

Once the solar potential of a home has been assessed, the next step is to consult with a reputable solar installer. A professional installer can provide a detailed site assessment, design a customized solar system, and guide homeowners through the installation process. It's important to choose an installer with experience and a track record of successful installations to ensure the system is designed and installed to maximize efficiency and performance.

In conclusion, assessing the solar potential of a home involves a comprehensive evaluation of geographic, structural, financial, and environmental factors. By carefully considering these elements, homeowners can make informed decisions about whether solar energy is a viable and beneficial option for their property. With the right approach and planning, solar energy can provide a sustainable and cost-effective solution for meeting residential energy needs, contributing to a cleaner and more sustainable future.

Installation and Maintenance of Solar Systems

Installing a solar system is a transformative step towards sustainable energy use, offering both environmental benefits and potential cost savings. However, the process requires careful planning and execution to ensure optimal performance and longevity. Understanding the intricacies of installation and maintenance is crucial for anyone considering this investment.

The installation process begins with a thorough site assessment. This involves evaluating the roof's condition, orientation, and shading, as well as determining the optimal placement for solar panels. A professional installer will conduct this assessment, taking into account factors such as roof pitch, structural integrity, and potential obstructions. It's essential to address any roof repairs or reinforcements before installation, as solar panels are designed to last for decades and should be supported by a stable foundation.

Once the site assessment is complete, the design phase begins. This involves selecting the appropriate type and number of solar panels, inverters, and mounting systems based on the site's characteristics and the homeowner's energy needs. The design should maximize energy production while adhering to local building codes and regulations. It's important to work with an experienced installer who can provide a customized solution tailored to the specific requirements of the property.

The actual installation process involves several key steps. First, the mounting system is securely attached to the roof, providing a stable base for the solar panels. This system must be installed with precision to ensure the panels are properly aligned and angled for maximum sunlight exposure. Next, the solar panels are mounted onto the system, followed by the installation of inverters and other electrical components. The inverters play a crucial role in converting the direct current (DC) generated by the panels into alternating current (AC) for use in the home.

Electrical wiring is a critical component of the installation process, requiring careful attention to detail and adherence to safety standards. The wiring connects the solar panels to the inverters and the home's electrical system, allowing for seamless integration and energy flow. A qualified electrician should perform this work to ensure compliance with local codes and to minimize the risk of electrical hazards.

Once the installation is complete, the system must be inspected and tested to ensure it is functioning correctly. This involves checking the alignment and connections of the panels, verifying the performance of the inverters, and ensuring the system is safely integrated with the home's electrical infrastructure. Many jurisdictions require a final inspection by a local authority or utility company before the system can be activated.

Maintenance is a vital aspect of ensuring the long-term performance and efficiency of a solar system. While solar panels are designed to be low-maintenance, regular inspections and cleaning are necessary to keep them operating at peak efficiency. Dust, dirt, and debris can accumulate on the panels, reducing their ability to capture sunlight. In areas with frequent rainfall, natural cleaning may suffice, but in drier regions, manual cleaning may be necessary. It's important to use appropriate cleaning methods and materials to avoid damaging the panels.

In addition to cleaning, periodic inspections should be conducted to check for any signs of wear or damage. This includes examining the mounting system, wiring, and inverters for any issues that could impact performance. Regular maintenance can help identify potential problems early, preventing costly repairs and ensuring the system continues to operate efficiently.

Monitoring the performance of a solar system is another important aspect of maintenance. Many modern systems come equipped with monitoring software that provides real-time data on energy production and usage. This allows homeowners to track the system's performance, identify any anomalies, and make informed decisions about maintenance or upgrades. Monitoring can also help optimize energy consumption, allowing homeowners to adjust their usage patterns to maximize savings.

Inverter maintenance is particularly important, as inverters are a critical component of the solar system and can affect overall performance. Inverters typically have a shorter lifespan than solar panels and may need to be replaced or serviced during the life of the system. Regular checks and updates to the inverter's software can help ensure it operates efficiently and reliably.

For those living in areas with harsh weather conditions, additional precautions may be necessary to protect the solar system. This can include securing panels against high winds, ensuring proper drainage to prevent water damage, and using protective coatings to guard against corrosion. Taking these steps can help extend the life of the system and maintain its performance in challenging environments.

Working with a reputable installer and maintenance provider is essential for the successful installation and upkeep of a solar system. A professional with experience and a proven track record can provide valuable guidance and support throughout the process, from initial assessment to ongoing maintenance. It's important to choose a provider who offers comprehensive services and stands behind their work with warranties and guarantees.

In conclusion, the installation and maintenance of a solar system require careful planning, execution, and ongoing attention. By understanding the key steps involved and working with experienced professionals, homeowners can ensure their solar investment delivers maximum benefits and contributes to a sustainable energy future. With proper care and maintenance, a solar system can provide reliable, clean energy for decades, reducing reliance on fossil fuels and lowering energy costs.

Financial Incentives and Savings

Navigating the financial landscape of solar energy can be a rewarding endeavor, offering significant savings and incentives that make the transition to renewable energy more accessible. Understanding the various financial incentives available and how they contribute to overall savings is crucial for anyone considering a solar investment. By leveraging these opportunities, homeowners and businesses can reduce the upfront costs of solar installations and enhance their long-term financial benefits.

One of the most significant financial incentives for solar energy is the federal investment tax credit (ITC), available in many countries. This credit allows homeowners and businesses to deduct a substantial percentage of the cost of installing a solar energy system from their federal taxes. The ITC has been instrumental in driving the growth of the solar industry, making solar installations more affordable and attractive. It's important to note that the percentage of the credit may vary over time, so staying informed about current legislation is essential to maximize this benefit.

In addition to federal incentives, many states and local governments offer their own programs to encourage solar adoption. These can include rebates, grants, and tax credits that further reduce the cost of solar installations. State-specific incentives can vary widely, so it's crucial to research the options available in your area. Some states also offer performance-based incentives, which provide payments based on the amount of electricity generated by the solar system. These programs reward efficient energy production and can significantly enhance the financial returns of a solar investment.

Net metering is another valuable financial mechanism that can lead to substantial savings for solar system owners. This policy allows homeowners and businesses to sell excess electricity generated by their solar panels back to the grid, receiving credits on their utility bills. Net metering effectively turns the utility grid into a virtual battery, storing excess energy for later use. The specifics of net metering policies can vary by region, including the rate at which credits are applied and any caps on the amount of energy that can be credited. Understanding the net metering policies in your area is crucial to maximizing the financial benefits of your solar system.

Some utilities offer additional incentives to encourage solar adoption, such as time-of-use (TOU) rates. TOU rates vary the cost of electricity based on the time of day, with higher rates during peak demand periods and lower rates during off-peak times. By aligning energy consumption with these rates, solar system owners can optimize their savings. For example, using stored solar energy during peak periods can reduce reliance on expensive grid electricity, leading to significant cost reductions.

Financing options play a critical role in making solar energy accessible to a broader audience. Many homeowners and businesses may not have the upfront capital to purchase a solar system outright, but various financing solutions can bridge this gap. Solar loans, for instance, allow individuals to finance the cost of a solar installation over time, often with favorable interest rates. These loans can be structured to match the expected savings from the solar system, making them an attractive option for those looking to minimize initial expenses.

Leasing and power purchase agreements (PPAs) are alternative financing models that have gained popularity in recent years. Under a solar lease, a third-party company owns and maintains the solar system, and the homeowner pays a fixed monthly fee to use the electricity generated. PPAs operate similarly, but the homeowner pays for the electricity produced at a predetermined rate. Both options can provide immediate savings on energy bills without the need for a large upfront investment, although they may not offer the same long-term financial benefits as owning the system outright.

The environmental benefits of solar energy also translate into financial savings. By reducing reliance on fossil fuels, solar energy helps decrease greenhouse gas emissions and air pollution, contributing to a cleaner environment. This can lead to indirect financial benefits, such as improved public health and reduced healthcare costs associated with pollution-related illnesses. Additionally, as governments and businesses increasingly prioritize sustainability, properties with solar installations may see an increase in value, providing a potential return on investment when selling the property.

Energy independence is another financial advantage of solar energy. By generating their own electricity, homeowners and businesses can reduce their vulnerability to fluctuating energy prices and potential disruptions in the energy supply. This stability can lead to more predictable energy costs and long-term savings, particularly in regions with volatile energy markets.

The combination of financial incentives, savings, and environmental benefits makes solar energy an attractive investment for many. However, it's essential to conduct a thorough financial analysis before proceeding with a solar installation. This includes evaluating the total cost of the system, potential savings from reduced energy bills, and the value of available incentives. By carefully considering these factors, individuals and businesses can make informed decisions that align with their financial goals and sustainability objectives.

As the solar industry continues to evolve, new financial opportunities and innovations are likely to emerge. Staying informed about the latest developments and understanding how they impact the financial landscape of solar energy is crucial for maximizing the benefits of a solar investment. With the right approach and planning, solar energy can provide significant financial rewards while contributing to a more sustainable future.

Overcoming Common Challenges

Embarking on the journey to harness solar energy can be both exciting and daunting. While the benefits of solar power are clear, the path to achieving them is often fraught with challenges. Understanding these common obstacles and how to overcome them is essential for anyone considering a solar installation. By addressing these issues head-on, homeowners and businesses can ensure a smoother transition to renewable energy.

One of the most prevalent challenges is the initial cost of solar installation. Despite the long-term savings and financial incentives available, the upfront investment can be a significant barrier for many. To tackle this, it's important to explore various financing options that can alleviate the financial burden. Solar loans, leases, and power purchase agreements (PPAs) offer flexible solutions that allow individuals to spread the cost over time. Additionally, researching and taking advantage of available tax credits, rebates, and grants can significantly reduce the initial expense, making solar energy more accessible.

Another common hurdle is navigating the complex regulatory landscape. Solar installations are subject to a myriad of local, state, and federal regulations, which can be overwhelming for those unfamiliar with the process. To overcome this challenge, it's crucial to work with a reputable solar installer who is well-versed in the regulatory requirements of your area. These professionals can guide you through the permitting process, ensuring compliance with all necessary codes and standards. Staying informed about any changes in regulations and policies is also important, as they can impact the feasibility and financial benefits of your solar project.

Shading and site suitability present additional challenges for solar installations. Trees, buildings, and other obstructions can cast shadows on solar panels, reducing their efficiency and energy output. Conducting a thorough site assessment is essential to identify potential sources of shading and determine the best placement for solar panels. In some cases, trimming trees or selecting a different roof section may be necessary to optimize sunlight exposure. For properties with limited roof space or unsuitable orientation, ground-mounted systems or community solar programs can offer viable alternatives.

Weather variability is another factor that can impact solar energy production. While solar panels are designed to operate in a range of conditions, their efficiency can be affected by factors such as cloud cover, temperature, and precipitation. To mitigate the effects of weather variability, consider incorporating energy storage solutions, such as batteries, into your solar system. These systems store excess energy generated during sunny periods for use during times of low sunlight, ensuring a consistent power supply. Additionally, monitoring and maintenance can help identify and address any performance issues related to weather conditions.

Maintenance and upkeep are often overlooked challenges in solar energy adoption. While solar panels are relatively low-maintenance, regular inspections and cleaning are necessary to ensure optimal performance. Dust, dirt, and debris can accumulate on the panels, reducing their ability to capture sunlight. Establishing a routine maintenance schedule and working with a professional service provider can help keep your system operating at peak efficiency. It's also important to monitor the performance of your solar system regularly, using available software and tools to track energy production and identify any anomalies.

Grid connectivity and integration pose another set of challenges for solar system owners. Ensuring that your solar installation is properly connected to the grid and compliant with utility requirements is essential for maximizing the benefits of your system. This includes understanding net metering policies, which allow you to sell excess electricity back to the grid, and any associated fees or restrictions. Working with an experienced installer who understands the intricacies of grid integration can help navigate these challenges and ensure a seamless connection.

Public perception and misinformation can also be obstacles to solar adoption. Despite the growing popularity of solar energy, misconceptions about its reliability, cost, and environmental impact persist. To overcome these challenges, it's important to educate yourself and others about the benefits and realities of solar energy. Engaging with community programs, attending informational sessions, and sharing your own experiences can help dispel myths and promote a more accurate understanding of solar technology.

Finally, technological advancements and market dynamics can present both challenges and opportunities for solar energy adoption. The rapid pace of innovation in the solar industry means that new technologies and products are constantly emerging, which can make it difficult to choose the best solution for your needs. Staying informed about the latest developments and working with knowledgeable professionals can help you make informed decisions and take advantage of new opportunities as they arise.

In conclusion, while the path to solar energy adoption is not without its challenges, these obstacles can be overcome with careful planning, research, and collaboration. By understanding the common challenges associated with solar installations and implementing strategies to address them, homeowners and businesses can successfully transition to renewable energy and enjoy the numerous benefits it offers. With the right approach, solar energy can provide a sustainable and cost-effective solution for meeting energy needs, contributing to a cleaner and more sustainable future.

Real-Life Examples of Residential Solar Success

The transformative power of solar energy is best illustrated through real-life examples of homeowners who have successfully integrated solar systems into their residences. These stories not only highlight the practical benefits of solar energy but also demonstrate the diverse ways in which individuals can tailor solar solutions to meet their unique needs and circumstances.

Consider the case of the Johnson family, who reside in a suburban neighborhood in California. Faced with rising electricity bills and a desire to reduce their environmental impact, the Johnsons decided to explore solar energy as a viable alternative. After conducting thorough research and consulting with a reputable solar installer, they opted for a rooftop solar system that would cover the majority of their energy needs. The installation process was seamless, and within a few months, the Johnsons began to see a significant reduction in their utility bills. By taking advantage of state and federal incentives, they were able to offset a substantial portion of the installation costs, making the investment financially feasible. The Johnsons' experience underscores the importance of leveraging available incentives and working with knowledgeable professionals to maximize the benefits of solar energy.

In another inspiring example, the Patel family in Texas faced a different set of challenges. Living in a region prone to extreme weather conditions, they were concerned about the reliability of their energy supply. To address this, they decided to install a solar-plus-storage system, which included a battery backup to store excess energy generated during sunny periods. This setup provided the Patels with a reliable power source during outages and peak demand times, ensuring their home remained powered even when the grid was down. The Patels' story highlights the versatility of solar energy solutions and the importance of customizing systems to address specific needs and concerns.

The story of Maria Lopez, a single mother living in a modest home in Arizona, offers another perspective on residential solar success. Maria was initially hesitant about the upfront costs of solar installation, but after learning about a local community solar program, she discovered a more affordable path to solar energy. By participating in the program, Maria was able to purchase a share of a larger solar array located off-site, allowing her to benefit from solar energy without installing panels on her own roof. This innovative approach not only reduced her energy bills but also empowered her to contribute to a more sustainable future. Maria's experience demonstrates the potential of community solar programs to make renewable energy accessible to a wider audience, particularly those who may face financial or logistical barriers to traditional solar installations.

In the bustling city of New York, the Chen family faced the challenge of limited roof space in their urban townhouse. Determined to embrace solar energy, they explored alternative solutions and ultimately decided on a building-integrated photovoltaic (BIPV) system. By incorporating solar cells into their building's facade and windows, the Chens were able to generate electricity while maintaining the aesthetic appeal of their home. This innovative approach not only provided them with clean energy but also enhanced the value of their property. The Chen family's story illustrates the creative solutions available to urban dwellers seeking to harness solar energy, even in space-constrained environments.

The experience of the Thompson family in rural Colorado offers yet another example of solar success. Living in an area with abundant sunlight but limited access to the electrical grid, the Thompsons decided to pursue an off-grid solar system. By combining solar panels with a robust battery storage system and a backup generator, they achieved complete energy independence. This self-sufficient setup allowed them to power their home sustainably while avoiding the high costs associated with extending the grid to their remote location. The Thompsons' journey underscores the potential of solar energy to provide reliable power in off-grid settings, offering a viable solution for those living in remote or underserved areas.

These real-life examples of residential solar success demonstrate the diverse ways in which solar energy can be adapted to meet the unique needs of different households. Whether through traditional rooftop installations, community solar programs, or innovative building-integrated solutions, solar energy offers a flexible and sustainable path to energy independence. By learning from these stories and understanding the various options available, homeowners can make informed decisions about how best to incorporate solar energy into their lives.

The common thread among these success stories is the proactive approach taken by each household. By conducting thorough research, consulting with experts, and exploring available incentives and programs, these homeowners were able to overcome challenges and achieve their solar energy goals. Their experiences serve as a testament to the transformative potential of solar energy and the positive impact it can have on both individual households and the broader community.

As the solar industry continues to evolve and expand, the opportunities for residential solar success will only grow. By staying informed about the latest developments and innovations, homeowners can continue to find new and creative ways to harness the power of the sun. These real-life examples serve as both inspiration and a roadmap for those considering the transition to solar energy, offering valuable insights into the practical and financial benefits of this sustainable energy source.

Chapter 3: Commercial and Industrial Solar Applications

Solar Integration in Businesses

Businesses across the globe are increasingly recognizing the value of integrating solar energy into their operations. This shift is driven by a combination of environmental responsibility, economic incentives, and the desire for energy independence. By adopting solar power, businesses can significantly reduce their carbon footprint, lower energy costs, and enhance their brand image as leaders in sustainability.

The journey to solar integration begins with a comprehensive assessment of the business's energy needs and infrastructure. This involves evaluating current energy consumption patterns, identifying peak usage times, and determining the potential for solar energy to meet these demands. A detailed energy audit can provide valuable insights into areas where solar power can be most effectively utilized, allowing businesses to tailor their solar solutions to their specific requirements.

Once the energy assessment is complete, the next step is to explore the various solar technologies and configurations available. Businesses have a range of options, from traditional rooftop solar panels to more innovative solutions like solar carports and building-integrated photovoltaics (BIPV). The choice of technology will depend on factors such as available space, architectural considerations, and budget constraints. For businesses with limited roof space, ground-mounted systems or solar canopies can offer viable alternatives, maximizing solar potential without compromising on aesthetics or functionality.

Financial considerations play a crucial role in the decision-making process for solar integration. While the initial investment can be substantial, the long-term savings on energy bills and the availability of financial incentives make solar energy an attractive proposition. Businesses can take advantage of federal and state tax credits, rebates, and grants to offset installation costs. Additionally, innovative financing options such as solar leases, power purchase agreements (PPAs), and green loans provide flexible solutions that align with a company's financial strategy.

The installation process for commercial solar systems requires careful planning and coordination. Working with experienced solar installers is essential to ensure that the system is designed and implemented to meet the business's energy goals. This includes obtaining the necessary permits, adhering to local building codes, and coordinating with utility companies for grid connection and net metering arrangements. A well-executed installation not only maximizes energy production but also minimizes disruptions to business operations.

Once the solar system is operational, ongoing monitoring and maintenance are key to ensuring optimal performance. Many commercial solar systems come equipped with advanced monitoring software that provides real-time data on energy production and consumption. This allows businesses to track the system's performance, identify any issues, and make informed decisions about energy management. Regular maintenance, including cleaning and inspections, is also important to keep the system running efficiently and to extend its lifespan.

Beyond the direct financial and environmental benefits, solar integration offers businesses a range of strategic advantages. By reducing reliance on fossil fuels, companies can mitigate the risks associated with volatile energy prices and supply disruptions. This energy independence provides greater stability and predictability in operational costs, allowing businesses to allocate resources more effectively.

Moreover, adopting solar energy can enhance a company's brand image and reputation. As consumers and stakeholders increasingly prioritize sustainability, businesses that demonstrate a commitment to renewable energy can differentiate themselves in the marketplace. This can lead to increased customer loyalty, improved employee satisfaction, and a stronger competitive position. By showcasing their solar initiatives, companies can also attract environmentally conscious investors and partners, further supporting their growth and success.

The integration of solar energy into business operations can also drive innovation and collaboration. By embracing renewable energy, companies can explore new business models, products, and services that align with their sustainability goals. This can include developing energy-efficient technologies, offering green products, or partnering with other organizations to create shared solar solutions. These initiatives not only contribute to a more sustainable future but also open up new revenue streams and opportunities for growth.

Real-life examples of successful solar integration in businesses abound, illustrating the diverse ways in which companies can harness the power of the sun. For instance, a manufacturing facility in the Midwest installed a large-scale rooftop solar array, reducing its energy costs by 30% and achieving a return on investment within five years. A retail chain in California implemented solar carports across its locations, providing shade for customers while generating clean energy to power its stores. These examples demonstrate the tangible benefits of solar energy and the potential for businesses to lead the way in the transition to a sustainable energy future.

In conclusion, solar integration in businesses offers a compelling opportunity to achieve environmental, economic, and strategic goals. By carefully assessing their energy needs, exploring available technologies, and leveraging financial incentives, companies can successfully incorporate solar power into their operations. The benefits extend beyond cost savings and environmental impact, positioning businesses as leaders in sustainability and innovation. As the solar industry continues to evolve, the potential for businesses to harness the power of the sun will only grow, paving the way for a brighter and more sustainable future.

Economic Advantages for Companies

Harnessing solar energy presents a multitude of economic advantages for companies, transforming the way businesses approach energy consumption and sustainability. By investing in solar power, companies can significantly reduce operational costs, enhance their financial stability, and position themselves as leaders in environmental responsibility. Understanding these economic benefits is crucial for businesses looking to make informed decisions about their energy strategies.

One of the most compelling economic advantages of solar energy is the potential for substantial cost savings on electricity bills. Traditional energy sources are subject to price volatility, which can lead to unpredictable and often rising energy costs. Solar energy, on the other hand, provides a stable and predictable source of power. By generating their own electricity, companies can shield themselves from fluctuating energy prices and reduce their reliance on the grid. This energy independence not only results in immediate savings but also provides long-term financial stability, allowing businesses to allocate resources more effectively.

The initial investment in solar technology can be offset by a variety of financial incentives, making the transition to solar energy more economically feasible. Governments at the federal, state, and local levels offer tax credits, rebates, and grants to encourage businesses to adopt renewable energy. These incentives can cover a significant portion of the installation costs, reducing the financial burden on companies. Additionally, accelerated depreciation programs, such as the Modified Accelerated Cost Recovery System (MACRS) in the United States, allow businesses to recover their solar investment more quickly by providing tax deductions over a shorter period.

Innovative financing options further enhance the economic appeal of solar energy for companies. Power purchase agreements (PPAs) and solar leases enable businesses to adopt solar technology with little to no upfront costs. Under a PPA, a third-party provider installs and maintains the solar system, and the company agrees to purchase the generated electricity at a predetermined rate. This arrangement allows businesses to benefit from solar energy without the financial responsibility of ownership. Similarly, solar leases provide companies with access to solar power in exchange for a fixed monthly payment, offering predictable energy costs and immediate savings.

Beyond direct financial savings, solar energy can enhance a company's brand image and competitive advantage. As consumers and investors increasingly prioritize sustainability, businesses that demonstrate a commitment to renewable energy can differentiate themselves in the marketplace. This positive brand association can lead to increased customer loyalty, attract environmentally conscious investors, and improve employee satisfaction. By showcasing their solar initiatives, companies can position themselves as forward-thinking leaders in sustainability, enhancing their reputation and market position.

Solar energy also offers companies the opportunity to generate additional revenue streams. Businesses with excess solar capacity can sell surplus electricity back to the grid through net metering programs, receiving credits or payments for the energy they contribute. This not only maximizes the return on investment but also supports the broader transition to renewable energy by contributing clean power to the community. Additionally, companies can explore partnerships with other organizations to develop shared solar projects, creating new business opportunities and fostering collaboration in the renewable energy sector.

The economic advantages of solar energy extend to risk management and resilience. By reducing reliance on fossil fuels, companies can mitigate the risks associated with energy supply disruptions and regulatory changes. This energy security provides greater operational stability and reduces vulnerability to external factors that could impact business continuity.

Furthermore, solar energy can enhance a company's resilience to climate-related risks, such as extreme weather events, by providing a reliable and sustainable power source.

Real-world examples of companies successfully leveraging solar energy illustrate the tangible economic benefits. A global technology firm, for instance, installed solar panels across its data centers, achieving significant energy cost reductions and enhancing its sustainability credentials. A manufacturing company in the Midwest integrated solar power into its operations, reducing its carbon footprint and achieving a return on investment within a few years. These examples demonstrate the diverse ways in which businesses can harness solar energy to achieve economic and environmental goals.

The integration of solar energy into business operations also drives innovation and efficiency. By adopting renewable energy, companies can explore new technologies and processes that align with their sustainability objectives. This can include developing energy-efficient products, optimizing supply chains, and implementing sustainable practices across the organization. These initiatives not only contribute to a more sustainable future but also enhance operational efficiency and competitiveness.

As the solar industry continues to evolve, the economic advantages for companies will only grow. Advances in solar technology, such as improved panel efficiency and energy storage solutions, will further enhance the financial benefits of solar energy. Additionally, as governments and businesses worldwide commit to reducing carbon emissions, the demand for renewable energy will continue to rise, creating new opportunities for companies to lead the way in the transition to a sustainable energy future.

In conclusion, the economic advantages of solar energy for companies are multifaceted and compelling. By reducing energy costs, leveraging financial incentives, and enhancing brand image, businesses can achieve significant financial and strategic benefits. The transition to solar energy not only supports environmental sustainability but also positions companies for long-term success in an increasingly competitive and environmentally conscious marketplace. As the solar industry continues to advance, the potential for businesses to capitalize on these economic advantages will only expand, paving the way for a brighter and more sustainable future.

Case Studies of Industrial Solar Projects

Industrial solar projects have emerged as a powerful force in the global shift towards renewable energy, offering substantial benefits to large-scale operations. These projects not only reduce carbon footprints but also provide significant cost savings and energy security. By examining real-world case studies, we can gain valuable insights into the successful implementation of solar energy in industrial settings and the transformative impact it can have on businesses.

One notable example is the solar initiative undertaken by a major automotive manufacturer in Germany. Faced with the dual challenge of rising energy costs and stringent environmental regulations, the company sought to integrate solar energy into its production facilities. The project involved the installation of a vast solar array on the rooftops of its manufacturing plants, covering an area equivalent to several football fields. This ambitious undertaking required meticulous planning and coordination, as the company had to navigate complex regulatory requirements and logistical challenges.

The results of the project were impressive. The solar installation generated a significant portion of the facility's energy needs, reducing reliance on the grid and lowering electricity costs. The company also benefited from government incentives, which helped offset the initial investment. Beyond the financial advantages, the project enhanced the company's reputation as a leader in sustainability, attracting environmentally conscious consumers and investors. This case study highlights the potential for solar energy to drive both economic and environmental benefits in the industrial sector.

In another compelling example, a large-scale agricultural operation in California embraced solar energy to address its energy-intensive processes. The farm, which specializes in the cultivation of high-value crops, faced escalating energy costs due to its reliance on traditional power sources for irrigation, processing, and storage. To mitigate these costs and reduce its environmental impact, the farm invested in a solar power system that included both rooftop panels and ground-mounted arrays.

The solar project was tailored to meet the farm's specific energy needs, taking into account factors such as seasonal variations in energy consumption and the availability of sunlight. By generating its own electricity, the farm achieved substantial cost savings and improved its energy independence. The project also provided a hedge against future energy price increases, enhancing the farm's financial stability. Additionally, the farm's commitment to renewable energy resonated with consumers, boosting its brand image and marketability. This case study demonstrates the versatility of solar energy solutions in addressing the unique challenges faced by different industries.

A third example comes from the mining sector, where a leading mining company in Australia implemented a solar-diesel hybrid system to power its remote operations. The mine, located in an area with limited access to the electrical grid, relied heavily on diesel generators for its energy needs. This dependence on fossil fuels resulted in high operational costs and significant greenhouse gas emissions. To address these issues, the company integrated solar power into its energy mix, creating a hybrid system that combined solar panels with existing diesel generators.

The hybrid system provided a reliable and sustainable energy solution, reducing diesel consumption and lowering emissions. The solar component of the system generated electricity during daylight hours, while the diesel generators provided backup power when needed. This approach not only reduced fuel costs but also minimized the environmental impact of the mining operations. The success of the project demonstrated the potential for solar energy to enhance the sustainability and efficiency of industrial operations in remote locations.

These case studies illustrate the diverse applications of solar energy in industrial settings and the significant benefits it can deliver. By adopting solar power, industrial operations can achieve cost savings, reduce their environmental impact, and enhance their energy security. The successful implementation of these projects requires careful planning, collaboration with experienced solar providers, and a commitment to sustainability.

The lessons learned from these case studies can inform future industrial solar projects, guiding businesses in their pursuit of renewable energy solutions. Key considerations include conducting a thorough energy assessment to understand the specific needs and challenges of the operation, exploring available financial incentives and financing options, and selecting the appropriate solar technology and configuration for the site.

Moreover, the integration of solar energy into industrial operations can drive innovation and efficiency. By embracing renewable energy, companies can explore new technologies and processes that align with their sustainability objectives. This can include optimizing production processes, developing energy-efficient products, and implementing sustainable practices across the organization. These initiatives not only contribute to a more sustainable future but also enhance operational efficiency and competitiveness.

As the solar industry continues to evolve, the potential for industrial solar projects will only grow. Advances in solar technology, such as improved panel efficiency and energy storage solutions, will further enhance the financial and environmental benefits of solar energy. Additionally, as governments and businesses worldwide commit to reducing carbon emissions, the demand for renewable energy will continue to rise, creating new opportunities for industrial operations to lead the way in the transition to a sustainable energy future.

In conclusion, the case studies of industrial solar projects demonstrate the transformative impact of solar energy on large-scale operations. By reducing energy costs, lowering emissions, and enhancing energy security, solar power offers a compelling solution for industrial businesses seeking to achieve their economic and environmental goals. The successful implementation of these projects serves as a testament to the potential of solar energy to drive positive change in the industrial sector, paving the way for a brighter and more sustainable future.

Overcoming Barriers in Commercial Adoption

Commercial adoption of solar energy presents a promising opportunity for businesses to reduce costs, enhance sustainability, and gain a competitive edge. However, the path to integrating solar power into commercial operations is not without its challenges. Understanding and overcoming these barriers is essential for businesses seeking to harness the benefits of solar energy.

One of the primary obstacles to commercial solar adoption is the perceived high upfront cost of installation. Many businesses are deterred by the initial investment required for solar panels and related infrastructure. To address this concern, it's crucial to explore the various financial incentives and financing options available. Governments often provide tax credits, rebates, and grants to encourage the adoption of renewable energy. These incentives can significantly reduce the initial financial burden, making solar energy more accessible to businesses. Additionally, innovative financing models such as power purchase agreements (PPAs) and solar leases allow companies to adopt solar technology with minimal upfront costs. By partnering with third-party providers, businesses can benefit from solar energy while paying a fixed rate for the electricity generated.

Another significant barrier is the complexity of navigating regulatory requirements and obtaining necessary permits. The process can be daunting, especially for businesses unfamiliar with the intricacies of solar energy regulations. To overcome this challenge, it's essential to work with experienced solar installers who are well-versed in local, state, and federal regulations. These professionals can guide businesses through the permitting process, ensuring compliance with all necessary codes and standards. Staying informed about any changes in regulations and policies is also important, as they can impact the feasibility and financial benefits of a solar project.

Space constraints and site suitability present additional challenges for commercial solar adoption. Not all businesses have the ideal roof space or orientation for solar panels, which can limit the potential for energy generation. Conducting a thorough site assessment is crucial to identify the best placement for solar panels and to explore alternative solutions. For businesses with limited roof space, ground-mounted systems or solar canopies can offer viable alternatives. Additionally, building-integrated photovoltaics (BIPV) provide an innovative solution by incorporating solar cells into the building's facade or windows, maximizing energy generation without compromising aesthetics.

The variability of solar energy production due to weather conditions is another concern for businesses considering solar adoption. While solar panels are designed to operate in a range of conditions, their efficiency can be affected by factors such as cloud cover, temperature, and precipitation. To mitigate the effects of weather variability, businesses can incorporate energy storage solutions, such as batteries, into their solar systems. These systems store excess energy generated during sunny periods for use during times of low sunlight, ensuring a consistent power supply. Additionally, integrating solar energy with other renewable sources, such as wind or biomass, can provide a more stable and reliable energy mix.

Maintenance and upkeep are often overlooked challenges in commercial solar adoption. While solar panels are relatively low-maintenance, regular inspections and cleaning are necessary to ensure optimal performance. Dust, dirt, and debris can accumulate on the panels, reducing their ability to capture sunlight. Establishing a routine maintenance schedule and working with a professional service provider can help keep the system operating at peak efficiency. It's also important to monitor the performance of the solar system regularly, using available software and tools to track energy production and identify any anomalies.

Grid connectivity and integration pose another set of challenges for businesses adopting solar energy. Ensuring that the solar installation is properly connected to the grid and compliant with utility requirements is essential for maximizing the benefits of the system. This includes understanding net metering policies, which allow businesses to sell excess electricity back to the grid, and any associated fees or restrictions. Working with an experienced installer who understands the intricacies of grid integration can help navigate these challenges and ensure a seamless connection.

Public perception and misinformation can also be obstacles to commercial solar adoption. Despite the growing popularity of solar energy, misconceptions about its reliability, cost, and environmental impact persist. To overcome these challenges, it's important for businesses to educate themselves and others about the benefits and realities of solar energy. Engaging with community programs, attending informational sessions, and sharing experiences can help dispel myths and promote a more accurate understanding of solar technology.

Finally, technological advancements and market dynamics can present both challenges and opportunities for commercial solar adoption. The rapid pace of innovation in the solar industry means that new technologies and products are constantly emerging, which can make it difficult for businesses to choose the best solution for their needs. Staying informed about the latest developments and working with knowledgeable professionals can help businesses make informed decisions and take advantage of new opportunities as they arise.

In conclusion, while the path to commercial solar adoption is not without its challenges, these obstacles can be overcome with careful planning, research, and collaboration. By understanding the common barriers associated with solar installations and implementing strategies to address them, businesses can successfully transition to renewable energy and enjoy the numerous benefits it offers. With the right approach, solar energy can provide a sustainable and cost-effective solution for meeting energy needs, contributing to a cleaner and more sustainable future.

Solar Solutions for Agricultural Needs

Agriculture, a cornerstone of human civilization, is increasingly turning to solar energy to address its unique challenges and enhance sustainability. The integration of solar solutions in agriculture not only reduces reliance on traditional energy sources but also offers innovative ways to improve efficiency and productivity. By exploring the diverse applications of solar energy in agriculture, we can uncover practical strategies that cater to the sector's specific needs.

One of the most significant applications of solar energy in agriculture is in irrigation systems. Water management is crucial for crop production, and traditional irrigation methods often rely on diesel or electric pumps, which can be costly and environmentally damaging. Solar-powered irrigation systems provide a sustainable alternative, harnessing the sun's energy to pump water from wells, rivers, or reservoirs. These systems are particularly beneficial in remote or off-grid areas where access to electricity is limited. By utilizing solar energy, farmers can reduce fuel costs, decrease greenhouse gas emissions, and ensure a reliable water supply for their crops.

Greenhouses, essential for extending growing seasons and protecting crops from adverse weather, can also benefit from solar energy. Solar panels can be installed on greenhouse roofs to generate electricity for heating, cooling, and lighting systems. This integration not only reduces energy costs but also enhances the environmental sustainability of greenhouse operations. Additionally, solar thermal systems can be used to heat water for irrigation or climate control, further optimizing resource use. By adopting solar solutions, greenhouse operators can improve their energy efficiency and reduce their carbon footprint.

Solar energy also plays a vital role in powering agricultural machinery and equipment. Tractors, harvesters, and other machinery traditionally rely on fossil fuels, contributing to pollution and operational costs. Solar-powered alternatives, such as electric tractors and machinery, offer a cleaner and more cost-effective solution. These machines can be charged using solar panels, reducing fuel expenses and emissions. Furthermore, portable solar chargers can be used to power smaller equipment and tools, providing flexibility and convenience for farmers working in the field.

Post-harvest processing and storage are critical components of the agricultural supply chain, and solar energy can significantly enhance these processes. Solar dryers, for example, offer an efficient way to dry crops such as grains, fruits, and vegetables, reducing spoilage and improving product quality. By using solar energy to remove moisture, farmers can preserve their produce without relying on electricity or fuel-powered drying methods. Similarly, solar-powered cold storage units provide a sustainable solution for preserving perishable goods, extending shelf life, and reducing food waste. These innovations not only improve the efficiency of post-harvest operations but also contribute to food security and economic stability.

Livestock farming can also benefit from solar energy solutions. Solar panels can power electric fences, water pumps, and lighting systems for barns and stables, reducing energy costs and enhancing animal welfare. Additionally, solar-powered ventilation systems can improve air quality and temperature control in livestock facilities, promoting healthier environments for animals. By integrating solar energy into livestock operations, farmers can achieve greater efficiency and sustainability while reducing their environmental impact.

The adoption of solar energy in agriculture is not without its challenges. Initial costs, technical expertise, and maintenance requirements can pose barriers for some farmers. However, various financial incentives and support programs are available to facilitate the transition to solar energy. Governments and organizations often offer grants, subsidies, and low-interest loans to help farmers invest in solar technology. Additionally, partnerships with solar providers and agricultural cooperatives can provide valuable resources and expertise, ensuring successful implementation and operation of solar systems.

Education and training are essential components of successful solar integration in agriculture. Farmers must be equipped with the knowledge and skills to operate and maintain solar systems effectively. Workshops, seminars, and online resources can provide valuable information on the benefits and applications of solar energy in agriculture. By fostering a culture of learning and innovation, the agricultural sector can embrace solar solutions and drive sustainable development.

Real-world examples of successful solar integration in agriculture highlight the transformative potential of renewable energy. In India, a solar-powered irrigation project has enabled smallholder farmers to access affordable and reliable water for their crops, improving yields and livelihoods. In Kenya, solar dryers have helped farmers reduce post-harvest losses and increase income by preserving fruits and vegetables for sale in local and international markets. These case studies demonstrate the diverse ways in which solar energy can address agricultural challenges and contribute to economic and environmental sustainability.

The future of solar energy in agriculture is promising, with ongoing advancements in technology and increasing awareness of the benefits of renewable energy. Innovations such as agrivoltaics, which combine solar panels with crop production, offer exciting opportunities for maximizing land use and energy generation. By strategically placing solar panels above crops, farmers can generate electricity while providing shade and reducing water evaporation, creating a symbiotic relationship between energy and agriculture.

In conclusion, solar solutions offer a powerful tool for addressing the unique needs of the agricultural sector. By harnessing the sun's energy, farmers can improve efficiency, reduce costs, and enhance sustainability across various aspects of their operations. The successful integration of solar energy in agriculture requires careful planning, collaboration, and education, but the potential benefits are substantial. As the world continues to seek sustainable solutions to global challenges, solar energy stands out as a beacon of hope for the future of agriculture.

Economic and Environmental Benefits

Solar energy offers a compelling array of economic and environmental benefits, making it an attractive option for businesses, governments, and individuals alike. As the world grapples with the dual challenges of climate change and energy security, the adoption of solar power presents a viable solution that addresses both economic and environmental concerns. By examining these benefits, we can better understand the transformative potential of solar energy and its role in shaping a sustainable future.

One of the most significant economic advantages of solar energy is the potential for substantial cost savings. Traditional energy sources, such as coal, oil, and natural gas, are subject to price volatility and supply constraints, leading to unpredictable and often rising energy costs. In contrast, solar energy provides a stable and predictable source of power. By generating their own electricity, businesses and households can reduce their reliance on the grid and shield themselves from fluctuating energy prices. This energy independence results in immediate savings on electricity bills and provides long-term financial stability, allowing for more effective resource allocation.

The initial investment in solar technology can be offset by a variety of financial incentives, making the transition to solar energy more economically feasible. Governments at various levels offer tax credits, rebates, and grants to encourage the adoption of renewable energy. These incentives can cover a significant portion of the installation costs, reducing the financial burden on businesses and individuals. Additionally, accelerated depreciation programs allow businesses to recover their solar investment more quickly by providing tax deductions over a shorter period. Innovative financing options, such as power purchase agreements (PPAs) and solar leases, further enhance the economic appeal of solar energy by enabling adoption with little to no upfront costs.

Beyond direct financial savings, solar energy can enhance brand image and competitive advantage. As consumers and investors increasingly prioritize sustainability, businesses that demonstrate a commitment to renewable energy can differentiate themselves in the marketplace. This positive brand association can lead to increased customer loyalty, attract environmentally conscious investors, and improve employee satisfaction. By showcasing their solar initiatives, companies can position themselves as forward-thinking leaders in sustainability, enhancing their reputation and market position.

Solar energy also offers significant environmental benefits, contributing to the reduction of greenhouse gas emissions and the mitigation of climate change. Unlike fossil fuels, solar power generates electricity without releasing harmful pollutants into the atmosphere. By transitioning to solar energy, businesses and households can significantly reduce their carbon footprint and contribute to cleaner air and water. This shift not only supports global efforts to combat climate change but also improves public health by reducing air pollution and its associated health risks.

The environmental benefits of solar energy extend to the conservation of natural resources. Traditional energy production often involves the extraction and consumption of finite resources, such as coal, oil, and natural gas. In contrast, solar energy harnesses the sun's abundant and renewable power, reducing the need for resource-intensive energy production. This sustainable approach helps preserve ecosystems and biodiversity, supporting a healthier planet for future generations.

Solar energy also plays a crucial role in enhancing energy security and resilience. By reducing reliance on imported fossil fuels, countries can achieve greater energy independence and stability. This energy security provides protection against geopolitical tensions and supply disruptions, ensuring a reliable and sustainable power supply. Additionally, solar energy systems can enhance resilience to climate-related risks, such as extreme weather events, by providing a decentralized and distributed energy source.

The integration of solar energy into various sectors can drive innovation and efficiency. By embracing renewable energy, businesses and governments can explore new technologies and processes that align with their sustainability objectives. This can include developing energy-efficient products, optimizing supply chains, and implementing sustainable practices across organizations. These initiatives not only contribute to a more sustainable future but also enhance operational efficiency and competitiveness.

Real-world examples of successful solar integration highlight the diverse applications and benefits of solar energy. In the United States, a major retailer installed solar panels across its stores, achieving significant energy cost reductions and enhancing its sustainability credentials. In India, a solar-powered irrigation project enabled smallholder farmers to access affordable and reliable water for their crops, improving yields and livelihoods. These case studies demonstrate the tangible benefits of solar energy and the potential for widespread adoption across different sectors and regions.

The future of solar energy is promising, with ongoing advancements in technology and increasing awareness of the benefits of renewable energy. Innovations such as improved panel efficiency, energy storage solutions, and agrivoltaics offer exciting opportunities for maximizing energy generation and land use. As governments and businesses worldwide commit to reducing carbon emissions, the demand for renewable energy will continue to rise, creating new opportunities for solar energy to lead the way in the transition to a sustainable energy future.

In conclusion, the economic and environmental benefits of solar energy are multifaceted and compelling. By reducing energy costs, lowering emissions, and enhancing energy security, solar power offers a powerful solution for addressing global challenges. The successful integration of solar energy requires careful planning, collaboration, and education, but the potential benefits are substantial. As the world continues to seek sustainable solutions, solar energy stands out as a beacon of hope for a cleaner, more sustainable future.

Case Studies of Solar-Powered Farms

Solar-powered farms are revolutionizing the agricultural landscape, offering a sustainable and efficient solution to the energy demands of modern farming. By harnessing the sun's energy, these farms not only reduce their carbon footprint but also enhance productivity and resilience. Examining real-world case studies provides valuable insights into the successful implementation of solar energy in agriculture and the transformative impact it can have on farming operations.

In the sun-drenched fields of California, a pioneering vineyard has embraced solar energy to power its operations. Faced with rising energy costs and a commitment to sustainability, the vineyard installed a solar array that spans several acres. This installation powers the entire winemaking process, from irrigation and refrigeration to bottling and distribution. The vineyard's solar system generates enough electricity to meet its energy needs, significantly reducing its reliance on the grid. The financial savings from lower energy bills have been substantial, allowing the vineyard to reinvest in its operations and enhance its sustainability initiatives. Moreover, the vineyard's commitment to renewable energy has resonated with environmentally conscious consumers, boosting its brand image and marketability.

In India, a solar-powered farm cooperative has transformed the lives of smallholder farmers. The cooperative, located in a region with limited access to electricity, faced challenges in irrigating crops and preserving produce. By installing solar-powered water pumps and cold storage units, the cooperative has overcome these obstacles. The solar pumps provide a reliable and cost-effective solution for irrigation, enabling farmers to cultivate crops year-round and increase yields. The solar-powered cold storage units help preserve perishable goods, reducing post-harvest losses and improving food security. This innovative approach has not only improved the livelihoods of the farmers but also contributed to the region's economic development.

A dairy farm in the Netherlands offers another compelling example of solar integration. The farm, which produces milk and cheese for local markets, sought to reduce its environmental impact and energy costs. By installing solar panels on the roofs of its barns and milking parlors, the farm generates a significant portion of its electricity needs. The solar energy powers milking machines, refrigeration units, and lighting systems, reducing the farm's reliance on fossil fuels. The farm has also implemented energy-efficient practices, such as using solar thermal systems to heat water for cleaning and sanitation. These initiatives have resulted in substantial cost savings and a reduction in greenhouse gas emissions, enhancing the farm's sustainability and profitability.

In Australia, a large-scale solar-powered cattle station demonstrates the potential for renewable energy in livestock farming. The station, located in a remote area with limited access to the electrical grid, relied heavily on diesel generators for its energy needs. To address the high costs and environmental impact of diesel consumption, the station installed a solar-diesel hybrid system. The solar panels generate electricity during daylight hours, while the diesel generators provide backup power when needed. This hybrid approach has reduced fuel costs and emissions, improving the station's operational efficiency and sustainability. The success of the project has inspired other cattle stations in the region to explore solar energy solutions.

These case studies illustrate the diverse applications and benefits of solar energy in agriculture. By adopting solar power, farms can achieve cost savings, reduce their environmental impact, and enhance their energy security. The successful implementation of these projects requires careful planning, collaboration with experienced solar providers, and a commitment to sustainability.

The lessons learned from these case studies can inform future solar-powered farm projects, guiding farmers in their pursuit of renewable energy solutions. Key considerations include conducting a thorough energy assessment to understand the specific needs and challenges of the farm, exploring available financial incentives and financing options, and selecting the appropriate solar technology and configuration for the site.

Moreover, the integration of solar energy into farming operations can drive innovation and efficiency. By embracing renewable energy, farmers can explore new technologies and processes that align with their sustainability objectives. This can include optimizing irrigation systems, developing energy-efficient equipment, and implementing sustainable practices across the farm. These initiatives not only contribute to a more sustainable future but also enhance operational efficiency and competitiveness.

As the solar industry continues to evolve, the potential for solar-powered farms will only grow. Advances in solar technology, such as improved panel efficiency and energy storage solutions, will further enhance the financial and environmental benefits of solar energy. Additionally, as governments and businesses worldwide commit to reducing carbon emissions, the demand for renewable energy will continue to rise, creating new opportunities for farms to lead the way in the transition to a sustainable energy future.

In conclusion, the case studies of solar-powered farms demonstrate the transformative impact of solar energy on agriculture. By reducing energy costs, lowering emissions, and enhancing energy security, solar power offers a compelling solution for addressing the challenges faced by modern farming. The successful integration of solar energy requires careful planning, collaboration, and education, but the potential benefits are substantial. As the world continues to seek sustainable solutions to global challenges, solar energy stands out as a beacon of hope for the future of agriculture.

Innovations in Solar Agriculture

Solar agriculture is undergoing a remarkable transformation, driven by innovative technologies and practices that are reshaping the way we produce food and manage resources. These advancements are not only enhancing the efficiency and sustainability of agricultural operations but also opening new avenues for growth and development. By exploring the latest innovations in solar agriculture, we can gain a deeper understanding of how these technologies are revolutionizing the industry and paving the way for a more sustainable future.

One of the most exciting developments in solar agriculture is the concept of agrivoltaics, which involves the simultaneous use of land for both solar energy production and agriculture. This innovative approach maximizes land use by installing solar panels above crops, allowing farmers to generate electricity while continuing to cultivate their fields. The strategic placement of solar panels provides shade for crops, reducing water evaporation and protecting plants from extreme temperatures. This dual-use system not only increases land productivity but also enhances crop resilience to climate change. Agrivoltaics has shown promising results in various regions, with studies indicating improved crop yields and reduced water usage.

Another groundbreaking innovation is the integration of solar-powered sensors and automation systems in precision agriculture. These technologies enable farmers to monitor and manage their fields with unprecedented accuracy, optimizing resource use and improving crop health. Solar-powered sensors collect real-time data on soil moisture, temperature, and nutrient levels, providing valuable insights into crop conditions. This information allows farmers to make informed decisions about irrigation, fertilization, and pest control, reducing waste and enhancing productivity. Automation systems, powered by solar energy, can further streamline operations by automating tasks such as irrigation and harvesting, freeing up time and resources for other activities.

Solar desalination is another promising innovation that addresses the critical issue of water scarcity in agriculture. This technology uses solar energy to convert seawater or brackish water into fresh water, providing a sustainable solution for irrigation in arid regions. Solar desalination systems are particularly beneficial in coastal areas where freshwater resources are limited. By harnessing the sun's energy, these systems offer an environmentally friendly alternative to traditional desalination methods, which are often energy-intensive and costly. The availability of fresh water through solar desalination can significantly enhance agricultural productivity and food security in water-stressed regions.

The development of solar-powered greenhouses represents another leap forward in sustainable agriculture. These greenhouses utilize solar panels to generate electricity for heating, cooling, and lighting, creating a controlled environment for year-round crop production. Solar-powered greenhouses can be equipped with advanced climate control systems that optimize temperature, humidity, and light levels, ensuring optimal growing conditions for a wide range of crops. This innovation not only reduces energy costs but also minimizes the environmental impact of greenhouse operations. By enabling continuous production, solar-powered greenhouses contribute to food security and reduce the carbon footprint of agricultural practices.

In addition to these technological advancements, solar agriculture is benefiting from innovative business models and financing options that facilitate the adoption of solar solutions. Community solar projects, for example, allow farmers to collectively invest in solar installations, sharing the costs and benefits of renewable energy. This collaborative approach makes solar energy more accessible to small and medium-sized farms, which may lack the resources to invest in individual systems. Power purchase agreements (PPAs) and solar leases offer additional financing options, enabling farmers to adopt solar technology with minimal upfront costs. These models provide flexibility and financial stability, allowing farmers to focus on their core operations while benefiting from clean energy.

The integration of solar energy into agricultural supply chains is also driving innovation and efficiency. Solar-powered cold storage and transportation systems help preserve the quality and freshness of produce, reducing post-harvest losses and food waste. By maintaining optimal temperatures during storage and transit, these systems extend the shelf life of perishable goods, enhancing marketability and profitability. Solar energy can also power processing facilities, reducing energy costs and emissions associated with food production. These innovations contribute to a more sustainable and resilient food system, supporting the transition to a low-carbon economy.

The potential of solar agriculture is further amplified by ongoing research and development efforts aimed at improving solar technology and its applications in agriculture. Advances in solar panel efficiency, energy storage solutions, and smart grid integration are enhancing the performance and reliability of solar systems. Researchers are also exploring new materials and designs for solar panels that can better withstand harsh environmental conditions and integrate seamlessly with agricultural landscapes. These developments are expanding the possibilities for solar agriculture, enabling farmers to harness the full potential of renewable energy.

As solar agriculture continues to evolve, it is essential to foster collaboration and knowledge sharing among stakeholders to accelerate the adoption of innovative solutions. Partnerships between farmers, researchers, technology providers, and policymakers can facilitate the exchange of ideas and best practices, driving progress and innovation. Education and training programs can equip farmers with the skills and knowledge needed to implement and manage solar technologies effectively. By building a supportive ecosystem for solar agriculture, we can unlock new opportunities for growth and sustainability.

In conclusion, innovations in solar agriculture are transforming the industry, offering sustainable solutions to the challenges of food production and resource management. By harnessing the power of the sun, farmers can enhance productivity, reduce environmental impact, and improve resilience to climate change. The successful integration of solar energy into agriculture requires a commitment to innovation, collaboration, and education, but the potential benefits are substantial. As the world seeks to address pressing global challenges, solar agriculture stands out as a beacon of hope for a more sustainable and prosperous future.

Challenges and Opportunities

Navigating the landscape of solar energy adoption in agriculture presents a unique set of challenges and opportunities. As the world increasingly turns to renewable energy sources, the agricultural sector stands at the forefront of this transition, poised to reap significant benefits while also confronting various hurdles. Understanding these dynamics is crucial for farmers, policymakers, and stakeholders seeking to harness the full potential of solar energy in agriculture.

One of the primary challenges facing the adoption of solar energy in agriculture is the initial cost of installation. Solar panels, inverters, and related infrastructure require a substantial upfront investment, which can be a deterrent for many farmers, particularly smallholders with limited financial resources. However, this challenge is not insurmountable. Various financial incentives, such as tax credits, grants, and subsidies, are available to offset these costs and make solar energy more accessible. Additionally, innovative financing models, such as power purchase agreements (PPAs) and solar leases, allow farmers to adopt solar technology with minimal upfront expenses, paying for the energy generated over time.

Another significant challenge is the technical expertise required to install, operate, and maintain solar systems. Many farmers may lack the necessary knowledge and skills to effectively manage solar installations, leading to potential inefficiencies and underperformance. To address this issue, it is essential to provide education and training programs that equip farmers with the skills needed to implement and maintain solar technologies. Collaborating with experienced solar providers and agricultural cooperatives can also provide valuable support and resources, ensuring successful integration and operation of solar systems.

Space constraints and site suitability present additional challenges for solar energy adoption in agriculture. Not all farms have the ideal roof space or land orientation for solar panels, which can limit the potential for energy generation. Conducting a thorough site assessment is crucial to identify the best placement for solar panels and explore alternative solutions. For farms with limited space, ground-mounted systems or solar canopies can offer viable alternatives. Additionally, building-integrated photovoltaics (BIPV) provide an innovative solution by incorporating solar cells into the building's facade or windows, maximizing energy generation without compromising aesthetics.

The variability of solar energy production due to weather conditions is another concern for farmers considering solar adoption. While solar panels are designed to operate in a range of conditions, their efficiency can be affected by factors such as cloud cover, temperature, and precipitation. To mitigate the effects of weather variability, farmers can incorporate energy storage solutions, such as batteries, into their solar systems. These systems store excess energy generated during sunny periods for use during times of low sunlight, ensuring a consistent power supply. Additionally, integrating solar energy with other renewable sources, such as wind or biomass, can provide a more stable and reliable energy mix.

Despite these challenges, the opportunities presented by solar energy in agriculture are substantial. One of the most significant benefits is the potential for cost savings. By generating their own electricity, farmers can reduce their reliance on the grid and shield themselves from fluctuating energy prices. This energy independence results in immediate savings on electricity bills and provides long-term financial stability, allowing for more effective resource allocation. The financial savings from lower energy costs can be reinvested in farm operations, enhancing productivity and sustainability.

Solar energy also offers environmental benefits, contributing to the reduction of greenhouse gas emissions and the mitigation of climate change. Unlike fossil fuels, solar power generates electricity without releasing harmful pollutants into the atmosphere. By transitioning to solar energy, farms can significantly reduce their carbon footprint and contribute to cleaner air and water. This shift not only supports global efforts to combat climate change but also improves public health by reducing air pollution and its associated health risks.

The integration of solar energy into agricultural operations can drive innovation and efficiency. By embracing renewable energy, farmers can explore new technologies and processes that align with their sustainability objectives. This can include developing energy-efficient equipment, optimizing irrigation systems, and implementing sustainable practices across the farm. These initiatives not only contribute to a more sustainable future but also enhance operational efficiency and competitiveness.

Solar energy also plays a crucial role in enhancing energy security and resilience. By reducing reliance on imported fossil fuels, countries can achieve greater energy independence and stability. This energy security provides protection against geopolitical tensions and supply disruptions, ensuring a reliable and sustainable power supply. Additionally, solar energy systems can enhance resilience to climate-related risks, such as extreme weather events, by providing a decentralized and distributed energy source.

Real-world examples of successful solar integration in agriculture highlight the diverse applications and benefits of solar energy. In the United States, a major dairy farm installed solar panels across its facilities, achieving significant energy cost reductions and enhancing its sustainability credentials. In Kenya, a solar-powered irrigation project enabled smallholder farmers to access affordable and reliable water for their crops, improving yields and livelihoods. These case studies demonstrate the tangible benefits of solar energy and the potential for widespread adoption across different sectors and regions.

The future of solar energy in agriculture is promising, with ongoing advancements in technology and increasing awareness of the benefits of renewable energy. Innovations such as improved panel efficiency, energy storage solutions, and agrivoltaics offer exciting opportunities for maximizing energy generation and land use. As governments and businesses worldwide commit to reducing carbon emissions, the demand for renewable energy will continue to rise, creating new opportunities for farms to lead the way in the transition to a sustainable energy future.

In conclusion, while the path to solar energy adoption in agriculture is not without its challenges, these obstacles can be overcome with careful planning, collaboration, and education. By understanding the common barriers associated with solar installations and implementing strategies to address them, farmers can successfully transition to renewable energy and enjoy the numerous benefits it offers. With the right approach, solar energy can provide a sustainable and cost-effective solution for meeting energy needs, contributing to a cleaner and more sustainable future.

The Concept of Community Solar

Community solar represents a transformative approach to renewable energy, offering a collective solution that allows individuals and businesses to benefit from solar power without the need for individual installations. This innovative model democratizes access to solar energy, making it possible for people who may not have the means or suitable conditions for solar panels on their own properties to participate in the clean energy revolution. By pooling resources and sharing the benefits, community solar projects foster a sense of collective ownership and responsibility towards sustainable energy practices.

At its core, community solar involves a shared solar array, often referred to as a solar garden or farm, which is typically located off-site from the participants' residences or businesses. Participants, who can be homeowners, renters, or businesses, subscribe to a portion of the solar array's output. In return, they receive credits on their electricity bills for the energy produced by their share of the solar project. This model provides an accessible entry point for those who are unable to install solar panels due to financial constraints, unsuitable roof conditions, or rental agreements.

The concept of community solar is particularly appealing in urban areas where space is limited, and rooftops may not be conducive to solar installations. By utilizing larger, centralized solar arrays, community solar projects can achieve economies of scale, reducing the overall cost of solar energy production.

This cost-effectiveness is passed on to participants, who can enjoy the benefits of solar energy at a lower price point than individual installations might offer. Additionally, community solar projects can be strategically located to maximize solar exposure and efficiency, further enhancing their economic viability.

One of the key advantages of community solar is its inclusivity. It opens the door for a diverse range of participants, including low-income households, to access renewable energy. Many community solar programs are designed with equity in mind, offering special incentives or pricing structures to ensure that underserved communities can participate. This inclusivity not only broadens the reach of solar energy but also contributes to social equity by providing all individuals with the opportunity to reduce their carbon footprint and lower their energy costs.

The development of community solar projects often involves collaboration between various stakeholders, including local governments, utility companies, solar developers, and community organizations. This collaborative approach ensures that projects are tailored to meet the specific needs and goals of the community. Local governments can play a crucial role by providing support through policy frameworks, zoning regulations, and financial incentives. Utility companies, on the other hand, facilitate the integration of community solar into the existing energy grid, ensuring that participants receive the appropriate credits on their bills.

Community engagement is a vital component of successful community solar projects. By involving residents and local organizations in the planning and decision-making process, projects can better align with the values and priorities of the community. This engagement fosters a sense of ownership and pride among participants, strengthening the project's long-term sustainability. Educational initiatives and outreach programs can further enhance community involvement by raising awareness about the benefits of solar energy and encouraging broader participation.

The environmental benefits of community solar are significant. By increasing the adoption of solar energy, these projects contribute to the reduction of greenhouse gas emissions and the transition to a cleaner energy grid. Community solar also supports local environmental goals, such as improving air quality and reducing reliance on fossil fuels. By participating in community solar, individuals and businesses can actively contribute to these environmental objectives, making a tangible impact on their local ecosystem.

Economic benefits extend beyond individual participants to the broader community. Community solar projects create jobs in the renewable energy sector, from installation and maintenance to project management and customer service. These jobs often provide opportunities for local residents, contributing to economic development and resilience. Additionally, by reducing energy costs for participants, community solar projects can increase disposable income, which can be reinvested in the local economy.

Real-world examples of community solar projects highlight their potential to drive positive change. In Minnesota, a state with a robust community solar program, thousands of residents and businesses have subscribed to community solar gardens, collectively generating significant amounts of clean energy. These projects have not only reduced energy costs for participants but have also contributed to the state's renewable energy targets. In New York, community solar initiatives have focused on expanding access to low-income households, demonstrating the model's potential to address energy equity.

The future of community solar is promising, with growing interest and investment in this model. As technology advances and costs continue to decline, community solar projects are becoming increasingly feasible and attractive. Policymakers and industry leaders are recognizing the value of community solar as a tool for achieving renewable energy goals and addressing climate change. By continuing to support and expand community solar initiatives, we can accelerate the transition to a sustainable energy future.

In conclusion, community solar offers a powerful solution for expanding access to renewable energy and fostering community engagement in sustainability efforts. By pooling resources and sharing the benefits, community solar projects provide an inclusive and cost-effective pathway to solar energy adoption. The success of these projects depends on collaboration, community involvement, and supportive policy frameworks. As we continue to explore innovative approaches to renewable energy, community solar stands out as a beacon of hope for a more equitable and sustainable future.

Successful Community Solar Projects

Community solar projects have emerged as a beacon of innovation and collaboration, offering a sustainable energy solution that benefits both individuals and communities. These projects exemplify how collective efforts can lead to significant environmental and economic gains. By examining successful community solar initiatives, we can glean valuable insights into the factors that contribute to their success and the impact they have on local communities.

In the heart of Colorado, the town of Fort Collins embarked on a community solar project that has become a model for others to follow. Faced with the challenge of increasing energy demands and a commitment to sustainability, the local government partnered with a solar developer to create a solar garden that serves hundreds of residents and businesses. This project was designed with inclusivity in mind, offering subscriptions to a diverse range of participants, including low-income households. By providing affordable access to solar energy, the project has helped reduce energy costs for participants while contributing to the town's renewable energy goals. The success of the Fort Collins solar garden can be attributed to strong community engagement, supportive policies, and a clear vision for sustainability.

In Massachusetts, the town of Harvard launched a community solar initiative that has transformed its energy landscape. The project, developed in collaboration with a local energy cooperative, involved the installation of a solar array on a former landfill site. This innovative use of land not only provided a new purpose for an otherwise unused area but also generated clean energy for the community. The project was funded through a combination of grants, loans, and community investments, demonstrating the power of collective financial support. Participants in the Harvard community solar project receive credits on their electricity bills, resulting in significant cost savings. The project's success has inspired neighboring towns to explore similar initiatives, highlighting the potential for community solar to drive regional change.

In Minnesota, a state known for its robust community solar program, the city of Minneapolis has embraced solar energy through a series of community solar gardens. These projects have been instrumental in expanding access to renewable energy, particularly for renters and residents of multi-family housing who may not have the option to install solar panels on their properties. By subscribing to a share of a solar garden, participants benefit from reduced energy costs and contribute to the city's sustainability goals. The Minneapolis community solar projects have been praised for their transparency and community involvement, with regular updates and meetings to keep participants informed and engaged. This open communication has fostered trust and enthusiasm, ensuring the long-term success of the projects.

In New York, the town of Ithaca has taken a unique approach to community solar by integrating it with local food production. The Ithaca community solar farm is located on a working farm, where solar panels coexist with crops and livestock. This innovative model, known as agrivoltaics, maximizes land use and provides dual benefits of clean energy and agricultural production. The project has garnered attention for its creative use of space and its contribution to local food security. By participating in the Ithaca community solar farm, residents not only reduce their energy costs but also support sustainable agriculture. The project's success has sparked interest in agrivoltaics as a viable model for other rural communities seeking to balance energy production with agricultural needs.

In California, the city of San Diego has implemented a community solar project that focuses on energy equity and environmental justice. Recognizing the need to address disparities in energy access, the project prioritizes low-income neighborhoods and communities of color. By offering affordable subscriptions and targeted outreach, the San Diego community solar initiative has empowered underserved communities to participate in the clean energy transition. The project has also provided job training and employment opportunities in the solar industry, contributing to local economic development. The success of the San Diego community solar project underscores the importance of equity and inclusion in renewable energy initiatives, demonstrating that community solar can be a powerful tool for social change.

These successful community solar projects share several common elements that contribute to their effectiveness. Strong partnerships between local governments, solar developers, and community organizations are crucial for navigating the complexities of project development and ensuring alignment with community goals. Transparent communication and community engagement foster trust and buy-in, creating a sense of ownership among participants. Financial incentives and innovative funding models, such as grants, loans, and community investments, make solar energy accessible to a broader audience. Additionally, a focus on inclusivity and equity ensures that the benefits of solar energy are shared by all members of the community.

The impact of successful community solar projects extends beyond individual participants to the broader community and environment. By increasing the adoption of solar energy, these projects contribute to the reduction of greenhouse gas emissions and the transition to a cleaner energy grid. They also support local economic development by creating jobs and reducing energy costs, which can be reinvested in the community. Furthermore, community solar projects enhance energy resilience by diversifying the energy supply and reducing reliance on fossil fuels.

As the demand for renewable energy continues to grow, the potential for community solar projects to drive positive change is immense. By learning from successful initiatives and applying their lessons to new projects, communities can harness the power of solar energy to achieve their sustainability goals. The future of community solar is bright, with ongoing advancements in technology and increasing awareness of the benefits of renewable energy. By continuing to support and expand community solar initiatives, we can accelerate the transition to a sustainable energy future and create a more equitable and resilient society.

In conclusion, successful community solar projects demonstrate the transformative potential of collective action in the pursuit of renewable energy. By fostering collaboration, inclusivity, and innovation, these projects provide a blueprint for communities seeking to embrace solar energy and achieve their sustainability objectives. As we continue to explore new approaches to renewable energy, community solar stands out as a powerful and effective model for driving positive change and building a more sustainable future.

Economic and Social Benefits

Harnessing solar energy offers a multitude of economic and social benefits that extend far beyond the immediate reduction of electricity bills. As communities and individuals increasingly turn to solar power, they unlock a range of opportunities that contribute to economic growth, social equity, and environmental sustainability. Understanding these benefits is crucial for anyone considering the transition to solar energy, as it highlights the broader impact of renewable energy adoption.

One of the most significant economic benefits of solar energy is the potential for cost savings. By generating their own electricity, individuals and businesses can reduce their reliance on the grid and protect themselves from fluctuating energy prices. This energy independence translates into immediate savings on electricity bills, providing financial relief and stability. Over time, the cumulative savings can be substantial, allowing for reinvestment in other areas of personal or business development. For businesses, reduced energy costs can enhance competitiveness and profitability, while for households, the savings can increase disposable income and improve quality of life.

The solar industry is a powerful engine for job creation, offering employment opportunities across various sectors, from manufacturing and installation to maintenance and project management. As the demand for solar energy continues to grow, so too does the need for skilled workers to support this burgeoning industry. This demand creates jobs that often provide competitive wages and opportunities for career advancement. Moreover, the decentralized nature of solar energy means that jobs are created locally, contributing to regional economic development and resilience. By investing in solar energy, communities can stimulate local economies and reduce unemployment rates, fostering a more prosperous and stable society.

Solar energy also plays a crucial role in enhancing energy security and resilience. By reducing reliance on imported fossil fuels, countries can achieve greater energy independence and stability. This energy security provides protection against geopolitical tensions and supply disruptions, ensuring a reliable and sustainable power supply. Additionally, solar energy systems can enhance resilience to climate-related risks, such as extreme weather events, by providing a decentralized and distributed energy source. This resilience is particularly important for communities vulnerable to natural disasters, as it ensures continuity of power supply and supports recovery efforts.

The social benefits of solar energy are equally compelling, particularly in terms of promoting equity and inclusivity. Solar energy has the potential to democratize access to clean power, providing opportunities for all individuals, regardless of income or location, to participate in the renewable energy transition. Community solar projects, for example, allow individuals who may not have suitable conditions for solar panels on their own properties to benefit from solar energy. These projects often prioritize low-income households, offering affordable subscriptions and targeted outreach to ensure inclusivity. By expanding access to solar energy, communities can address energy poverty and reduce disparities in energy access, contributing to a more equitable society.

Education and awareness are critical components of the social benefits of solar energy. As communities adopt solar power, they often engage in educational initiatives that raise awareness about the benefits of renewable energy and encourage broader participation. These initiatives can take the form of workshops, seminars, and outreach programs that inform individuals about the environmental and economic advantages of solar energy. By fostering a culture of sustainability and environmental stewardship, these educational efforts contribute to long-term behavioral change and support the transition to a low-carbon economy.

The environmental benefits of solar energy are well-documented, with solar power playing a key role in reducing greenhouse gas emissions and mitigating climate change. Unlike fossil fuels, solar energy generates electricity without releasing harmful pollutants into the atmosphere. By transitioning to solar power, individuals and businesses can significantly reduce their carbon footprint and contribute to cleaner air and water. This shift not only supports global efforts to combat climate change but also improves public health by reducing air pollution and its associated health risks. The environmental benefits of solar energy extend to biodiversity conservation, as solar installations can be designed to coexist with natural habitats and support local ecosystems.

Real-world examples of the economic and social benefits of solar energy abound. In Germany, a country known for its commitment to renewable energy, the solar industry has created thousands of jobs and contributed to the country's economic growth. The widespread adoption of solar power has also reduced Germany's reliance on imported fossil fuels, enhancing energy security and resilience. In India, solar energy projects have brought electricity to remote villages, improving quality of life and supporting economic development. These projects have provided access to education, healthcare, and other essential services, demonstrating the transformative potential of solar energy in addressing social challenges.

The future of solar energy is bright, with ongoing advancements in technology and increasing awareness of the benefits of renewable energy. As solar panels become more efficient and affordable, the barriers to adoption continue to diminish, making solar power accessible to a broader audience. Policymakers and industry leaders are recognizing the value of solar energy as a tool for achieving economic growth, social equity, and environmental sustainability. By continuing to support and expand solar initiatives, we can accelerate the transition to a sustainable energy future and create a more equitable and resilient society.

In conclusion, the economic and social benefits of solar energy are vast and far-reaching, offering a compelling case for its adoption. By reducing energy costs, creating jobs, and promoting equity, solar power contributes to a more prosperous and sustainable future. The environmental benefits further underscore the importance of transitioning to renewable energy, as solar power plays a crucial role in combating climate change and protecting public health. As we continue to explore innovative approaches to renewable energy, solar power stands out as a powerful and effective model for driving positive change and building a more sustainable future.

Overcoming Challenges in Community Adoption

Community solar projects offer a promising pathway to renewable energy adoption, yet they are not without their challenges. Successfully implementing these projects requires overcoming a range of obstacles, from financial and logistical hurdles to social and regulatory barriers. Understanding these challenges and developing strategies to address them is essential for communities seeking to harness the benefits of solar energy.

One of the primary challenges in community solar adoption is securing the necessary funding and financial support. The initial costs of developing a solar project, including site acquisition, equipment, and installation, can be substantial. For many communities, especially those with limited financial resources, these costs can be a significant barrier. To overcome this challenge, communities can explore various funding options, such as grants, loans, and public-private partnerships. Engaging with local governments and financial institutions can provide access to resources and incentives that reduce the financial burden. Additionally, innovative financing models, such as community investment funds and crowdfunding, can mobilize local support and generate the capital needed to launch a project.

Regulatory and policy barriers also pose significant challenges to community solar adoption. Navigating the complex landscape of energy regulations and policies can be daunting, particularly for communities with limited experience in renewable energy projects. To address these challenges, it is crucial to engage with policymakers and regulatory bodies early in the project development process. Building relationships with key stakeholders and advocating for supportive policies can help create a favorable environment for community solar projects. Additionally, collaborating with experienced solar developers and legal experts can provide valuable guidance and ensure compliance with relevant regulations.

Site selection and land use considerations are critical factors in the success of community solar projects. Identifying suitable locations for solar installations can be challenging, particularly in densely populated urban areas where space is limited. To overcome this obstacle, communities can explore creative solutions, such as utilizing underutilized or vacant land, repurposing brownfield sites, or installing solar panels on rooftops and parking structures. Conducting thorough site assessments and feasibility studies can help identify the most viable locations and optimize the design of the solar array.

Community engagement and participation are essential components of successful community solar projects. Building trust and buy-in from local residents and stakeholders is crucial for overcoming social barriers and ensuring long-term project sustainability. To foster community support, it is important to engage residents in the planning and decision-making process, providing opportunities for input and feedback. Hosting informational meetings, workshops, and outreach events can raise awareness about the benefits of solar energy and encourage broader participation. Transparent communication and regular updates can further strengthen community trust and commitment to the project.

Addressing equity and inclusivity is another critical challenge in community solar adoption. Ensuring that all community members, regardless of income or background, have access to the benefits of solar energy is essential for promoting social equity. To achieve this, communities can design projects with equity in mind, offering affordable subscription options and targeted outreach to underserved populations. Partnering with local organizations and advocacy groups can help identify and address barriers to participation, ensuring that the project is inclusive and accessible to all.

Technical expertise and capacity building are also important considerations in overcoming challenges to community solar adoption. Many communities may lack the technical knowledge and skills needed to develop and manage a solar project. To address this gap, it is important to invest in education and training programs that equip community members with the necessary skills and knowledge. Collaborating with experienced solar providers and industry experts can provide valuable support and resources, ensuring successful project implementation and operation.

Real-world examples of communities that have successfully overcome challenges in solar adoption offer valuable lessons and inspiration. In the city of Portland, Oregon, a community solar project was developed on a former landfill site, transforming an underutilized area into a source of clean energy. The project was made possible through a combination of public and private funding, strong community engagement, and supportive policies. In Washington, D.C., a community solar initiative focused on expanding access to low-income households, offering affordable subscriptions and targeted outreach to ensure inclusivity. These projects demonstrate the power of collaboration, innovation, and perseverance in overcoming challenges and achieving successful solar adoption.

The future of community solar is bright, with ongoing advancements in technology and increasing awareness of the benefits of renewable energy. As solar panels become more efficient and affordable, the barriers to adoption continue to diminish, making solar power accessible to a broader audience. Policymakers and industry leaders are recognizing the value of community solar as a tool for achieving renewable energy goals and addressing climate change. By continuing to support and expand community solar initiatives, we can accelerate the transition to a sustainable energy future and create a more equitable and resilient society.

In conclusion, while the path to community solar adoption is not without its challenges, these obstacles can be overcome with careful planning, collaboration, and innovation. By understanding the common barriers associated with solar projects and implementing strategies to address them, communities can successfully transition to renewable energy and enjoy the numerous benefits it offers. With the right approach, community solar can provide a sustainable and cost-effective solution for meeting energy needs, contributing to a cleaner and more sustainable future.

Inspiring Stories of Community Collaboration

In the realm of renewable energy, community collaboration has proven to be a powerful catalyst for change. Across the globe, communities have come together to harness the power of solar energy, creating inspiring stories of collective action and innovation. These stories not only highlight the potential of solar energy but also demonstrate the strength of community spirit and the impact of working together towards a common goal.

In the small town of Greensburg, Kansas, a devastating tornado in 2007 left the community in ruins. Faced with the daunting task of rebuilding, the residents of Greensburg saw an opportunity to transform their town into a model of sustainability. Embracing renewable energy as a cornerstone of their recovery, the community embarked on an ambitious plan to become one of the greenest towns in America. Solar energy played a pivotal role in this transformation, with community members collaborating to install solar panels on public buildings, homes, and businesses. This collective effort not only reduced the town's carbon footprint but also fostered a sense of unity and resilience among residents. Today, Greensburg stands as a testament to the power of community collaboration and the potential of renewable energy to drive positive change.

In the bustling city of Copenhagen, Denmark, a unique community solar project has captured the imagination of residents and visitors alike. Known as the "Solar City Tower," this innovative project involves the installation of solar panels on the iconic Copenhagen Opera House. The project was initiated by a group of local artists and environmentalists who saw an opportunity to combine art and sustainability in a way that would engage the community. Through a series of workshops and public events, the group garnered support from residents, businesses, and the local government, ultimately securing the funding needed to bring their vision to life. The Solar City Tower not only generates clean energy for the opera house but also serves as a symbol of Copenhagen's commitment to sustainability and community collaboration.

In the rural village of Dharnai, India, a lack of access to reliable electricity had long hindered economic development and quality of life. Determined to change this, the villagers partnered with a non-governmental organization to develop a community solar microgrid. This innovative project involved the installation of solar panels and battery storage systems, providing the village with a reliable and sustainable source of electricity. The project was designed with community involvement at its core, with villagers participating in the planning, installation, and maintenance of the solar systems. This collaborative approach not only empowered the community but also ensured the project's long-term success. Today, Dharnai is a thriving village with access to electricity for homes, schools, and businesses, demonstrating the transformative power of community collaboration and solar energy.

In the heart of Brooklyn, New York, a grassroots initiative known as the Brooklyn Microgrid has redefined the concept of community energy. This innovative project allows residents to generate, store, and trade solar energy within their neighborhood, creating a decentralized and resilient energy system. The project was born out of a desire to increase energy independence and resilience in the face of climate change and natural disasters. Through a series of community meetings and workshops, residents came together to design and implement the microgrid, with support from local businesses and energy experts. The Brooklyn Microgrid not only provides clean energy to the community but also fosters a sense of empowerment and ownership among participants. This pioneering project has inspired similar initiatives in other urban areas, highlighting the potential of community collaboration to drive innovation in renewable energy.

In the coastal town of Byron Bay, Australia, a community-driven solar project has captured the hearts and minds of residents.

Known as the "Byron Bay Solar Train," this project involves the conversion of a historic train into a solar-powered passenger service. The idea was conceived by a group of local environmentalists and train enthusiasts who saw an opportunity to combine their passions for sustainability and heritage preservation. Through a series of fundraising events and community partnerships, the group raised the necessary funds to retrofit the train with solar panels and battery storage. The Byron Bay Solar Train now operates as a popular tourist attraction, showcasing the potential of solar energy and community collaboration. This project has not only reduced the town's carbon emissions but also strengthened community ties and inspired a new generation of environmental advocates.

These inspiring stories of community collaboration demonstrate the transformative potential of solar energy and the power of collective action. By working together towards a common goal, communities can overcome challenges, drive innovation, and create lasting change. The success of these projects is a testament to the strength of community spirit and the impact of collaboration in the pursuit of a sustainable future.

The lessons learned from these stories can serve as a blueprint for other communities seeking to harness the power of solar energy. Key elements of success include strong community engagement, innovative thinking, and a commitment to inclusivity and equity. By fostering a culture of collaboration and empowerment, communities can unlock the full potential of solar energy and create a more sustainable and equitable world.

As the demand for renewable energy continues to grow, the potential for community collaboration to drive positive change is immense. By learning from these inspiring stories and applying their lessons to new projects, communities can harness the power of solar energy to achieve their sustainability goals. The future of community solar is bright, with ongoing advancements in technology and increasing awareness of the benefits of renewable energy. By continuing to support and expand community solar initiatives, we can accelerate the transition to a sustainable energy future and create a more equitable and resilient society.